LP 12117187

Analytical
Gas Chromatography

Analytical
Gas Chromatography

WALTER JENNINGS

Department of Food Science and Technology
University of California
Davis, California

1987

ACADEMIC PRESS, INC.

Harcourt Brace Jovanovich, Publishers

Orlando   San Diego   New York   Austin
Boston   London   Sydney   Tokyo   Toronto

ACADEMIC PRESS, INC.
Orlando, Florida 32887

United Kingdom Edition published by
ACADEMIC PRESS INC. (LONDON) LTD.
24–28 Oval Road, London NW1 7DX

Library of Congress Cataloging in Publication Data

Jennings, Walter
    Analytical gas chromatography.

    Includes index.
    1. Gas chromatography.   2. Capillarity.   I. Title.
QD79.C45 J458   1987   543'.0896   86-28873
ISBN 0–12–384355–3 (alk. paper)

PRINTED IN THE UNITED STATES OF AMERICA

87 88 89 90   9 8 7 6 5 4 3 2 1

CONTENTS

CHAPTER 3
Sample Injection

CHAPTER 4
The Stationary Phase

CHAPTER 5
Variables in the Gas Chromatographic Process

CHAPTER 6
Column Selection, Installation, and Use

CHAPTER 7
Instrument Conversion and Adaptation

CHAPTER 8
Special Analytical Techniques

CHAPTER 9
Selected Applications

PREFACE

This book is intended as a free-standing "introduction to and guide through" the rapidly developing field of analytical gas chromatography. It began as a third revised edition of "Gas Chromatography with Glass Capillary Columns," but it soon became evident that a separate offering would be preferable. The availability of a wide range of fused silica columns with different dimensional characteristics and coated with a variety of stationary phases has catalyzed changes in every aspect of the analytical process, from preparation and introduction of the sample to chromatographic separation of the solutes to their detection. These developments make an entirely fresh approach desirable, but it seems unnecessary to reproduce a detailed iteration of subjects whose coverage in the second edition of the earlier title remains adequate and current.

When beginning the composition of an instructional text, an author is immediately faced with a critical decision relative to the level of presentation; the text can be designed as a research tool, as a laboratory aid, or as something intermediate. My decision was influenced by the following factors:

Most of my life has been spent as a university professor. I was introduced to gas chromatography in 1954 by Dr. Keene Dimmick of the USDA laboratory in Albany, California. Gas chromatography soon became my major research interest. Professor Kurt Grob in Zurich gave me my first glass capillary column in 1965, with advice on its construction and use; my research activities soon turned in this new direction.

A few visionaries at Hewlett Packard recognized the potential of capillary gas chromatography in the early 1970s, and I was invited to participate in several of their "basic" gas chromatography courses, to consult on certain aspects of instrumental design, and eventually to instruct a series of special short courses on capillary gas chromatography throughout the United States and Canada. The latter arrangement continued for several years, concurrent with and followed by the instruction of capillary courses sponsored by other instrument companies and by chromatography discussion groups. These activities expanded with a number of invitations to instruct in-house courses for specific industries.

During this period, one of my completing doctoral students suggested the manufacture of glass capillary columns as a commercial business. My contributions in the early stages of this endeavor were largely in the form of teaching courses, handling technical inquiries, and troubleshooting.

Hence, the approach taken in this offering has been shaped by (1) more than thirty years experience in university teaching and research in gas chromatography; (2) many hours spent in answering telephoned and written queries on all aspects of gas chromatographic analysis; and (3) the instruction of between 350 and 400 intensive short courses on this subject, to a total of some 12,000 practitioners, at points all over the world.

These activities would leave almost anyone convinced that communication between most of those practicing gas chromatography on the analytical firing line and many of our leading chromatographic scientists leaves much to be desired. This may relate to the immense numbers of users and the diversity of their backgrounds and interests. The vast majority of practicing chromatographers has had little or no formal training in the subject. A large number of these users employ the method as a means to an end, and the technique is so powerful that useful data can be generated even by untrained people misusing poorly designed equipment. To extend the utility of this book to the largest possible number of users, I have deliberately stressed simplicity, especially in explanations. While there are benefits to many readers, there is also a definite hazard to this approach: Albert Einstein once observed that

> "Everything should be made as simple as possible . . .
> but not simpler."

I hope I have succeeded in not overstepping that line.

Finally, a good book should be "readable"; it should flow smoothly, logically, and bestow on the reader that satisfying feeling that comes only as he or she gains a better understanding of a technical subject. I sincerely hope that this offering succeeds, not only in the attainment of the above goals, but most of all, in increasing the proportion of analysts who do "good" chromatography.

Walter Jennings
Davis, California

CHAPTER 1

INTRODUCTION

1.1 General Considerations

Toward the end of the nineteenth century, the Russian scientist Mikhail Tswett demonstrated the separation of plant pigment mixtures into colored zones by percolating extracts through a column packed with an adsorbant. In the early 1900s, he used the word ''chromatography'' to describe this system of ''color graphics'' [1,2]. The word ''chromatography'' is now used as a generic term for separation processes that subject the substances to be separated to differential partitioning between two phases; in most cases, one phase is stationary and the other is mobile.

A. J. P. Martin and R. L. M. Synge later explored liquid–solid chromatography in great detail and were both eventually awarded Nobel Prizes for their efforts. It was in his award address that Martin first suggested that it should also be possible to employ a gas as the mobile phase in chromatographic processes. It is doubtful that he or anyone else, either at that time or some years later when James and Martin [3] demonstrated the first gas chromatographic separations, envisioned the degree to which this technique would come to dominate our analytical procedures.

Ray suggested combining thermal conductivity detection with gas chromatography, and in 1954 [4] he published schematics and chromatograms that stimulated a number of workers, including this author, to enter what promised to be a new and exciting field. Subsequent progress has been due to the individual contributions of some hundreds of scientists. Many of these myriad contributions have been detailed elsewhere (e.g., [5,6]), but three milestone achievements

1

must be mentioned before proceeding to a discussion of modern analytical gas chromatography: Golay's invention of the open tubular column [7], Desty's glass-drawing machine [8], and the thin-walled fused silica column (Dandeneau and Zerrener [9]).

When a gas is employed as the mobile phase, either a liquid or a solid can be utilized as the stationary phase; these processes are most precisely termed ''gas–liquid partition chromatography'' and ''gas–solid partition chromatography,'' respectively. Although the latter sees some applications in fixed gas analysis (see later chapters), the former is employed in the separation of most organic compounds and will receive most attention here. Through popular usage, the process is now more generally termed simply ''gas chromatography'' and is often abbreviated to ''GC.''

1.2   A Simplistic Approach

In the process of gas chromatography, the stationary phase is usually in the form of a thin film and is confined to an elongated tube, the column. The two extremes in column types are packed columns and open tubular columns. In the former, the thin film of stationary phase is distributed over an ''inert'' granular support, while in the latter the stationary phase is supported as a thin coating on the inner surface of the column wall (wall-coated open tubular columns). Popular usage soon equated the terms ''open tubular'' and ''capillary'' in gas chromatography. Some authorities decried this trend and recalled Golay's observation that it was ''the openess of the column, not its capillary dimensions'' that gave it special properties. Recently, however, larger-diameter open tubular columns have reappeared and in their new state show great promise of finally replacing the packed column. These large-diameter open tubular columns are not true ''capillaries'' and are incapable of producing what has come to be regarded as ''capillary chromatography.'' It has now become necessary to differentiate types of open tubular columns: in this book the use of the word ''capillaries'' will be restricted to open tubular columns whose inner diameters are less than 0.35 mm.

Be it packed or open tubular, the column, which normally begins at the inlet of the gas chromatograph and terminates at its detector, is adjusted to some suitable temperature and continuously swept with the mobile gas phase (carrier gas). When a mixture of volatile components is placed on the inlet end of the column, individual molecules of each of the solutes in that sample are swept toward the detector whenever they enter the moving stream of carrier gas. The proportion of each molecular species that is in the mobile phase at any given time is a function of the vapor pressure of each solute: The molecules of the components that exhibit higher vapor pressures remain largely in the mobile phase; they are swept toward the detector more rapidly and are the first solutes eluted from the column.

Other solutes exhibit lower vapor pressures, either because they are higher-boiling or because they engage in interactions with the stationary phase that effectively reduce their vapor pressures. Individual molecules of these solutes venture into the mobile phase (carrier gas) less frequently, their concentrations in the mobile phase are lower, and they require longer periods of time to reach the detector; hence separation is achieved.

Readers who are just beginning their comprehension of this powerful technique may find it helpful to (incorrectly) visualize gas chromatography as a stepwise process and to begin by considering the separation of a simple two-component mixture containing, e.g., acetone (bp 57°C) and ether (bp 37°C). If a small amount of that mixture is introduced into a chromatographic column which is continuously swept with carrier gas and held at a temperature where each solute exhibits a suitable vapor pressure, both solutes will immediately partition between the moving gas phase and the immobile stationary phase. All other things being equal, the molecules of the lower-boiling ether that are dissolved in stationary phase will vaporize before (or more frequently than) the molecules of the higher-boiling acetone [10]. As they enter the mobile gas phase they are carried down the column and over virgin stationary phase, where they redissolve. A fraction of a second before the acetone molecules revaporize to be carried downstream again by the carrier gas, the ether molecules move again; hence the more volatile ether molecules continuously increase their lead over the less volatile acetone molecules, and separation is achieved.

Although this concept may prove helpful in visualizing that a multiplicity of vaporizations and re-solutions on the part of the individual solute molecules is one factor influencing the degree of separation efficiency, it must be stressed that the oversimplification has resulted in an inaccurate picture. For one thing, the chromatographic process is continuous and highly dynamic rather than being a series of discrete steps. At any point in time, some of the molecules of each solute are in the stationary phase and others are in the mobile phase; as the mobile phase moves over virgin stationary phase, some of the mobile-phase-entrained solute molecules dissolve in the stationary phase, while immediately behind the moving front an equivalent number of dissolved solute molecules vaporize into the mobile phase. Because ether and acetone exhibit different vapor pressures, the ratio

molecules in mobile phase/molecules in stationary phase

will be larger for the more volatile ether than for the less volatile acetone. Hence, ether will move through the column more rapidly, and at the conclusion of the process a "plug" of ether molecules dispersed in mobile phase (carrier gas) will emerge to the detector, followed by a second "plug" of acetone.

The above concepts are also helpful in emphasizing that the vapor pressure of the solute strongly influences its chromatographic behavior; solutes undergo no separation in the mobile phase, nor do they undergo separation in the stationary phase. Solute separation is dependent on the differences in solute volatilities, which influence the rates (or frequencies) of solute vaporizations and re-solutions; this differentiates solute concentrations in the stationary and mobile phases. Hence it is desirable to subject solutes to as many "vaporization steps" as possible (without having other adverse effects, *vide infra*), and this will require that they undergo an equal number of "re-solution steps." If the vapor pressures of the solutes are too high, they spend most (or all) of their transit time in the mobile phase and little (or no) separation is achieved; if the vapor pressures are too low, the solutes spend too long in the stationary phase, analysis times become disproportionately long, and sensitivity is also adversely affected (*vide infra*). Column temperature is one obvious method of influencing solute vapor pressures; another is through the choice of stationary phase. A "polar" stationary phase reduces the vapor pressures of polar solutes by means of (additional) solute–stationary phase interactions that may include hydrogen bonding and/or dipole–dipole interactions. These interrelationships are discussed in greater detail in Chapter 4.

1.3 Simplistic Comparisons of Packed and Open Tubular Columns

Most chromatographers recognize that the open tubular (or "capillary") column is capable of separations that are vastly superior to those obtained on packed columns [11]. Figure 1.1 illustrates separations of an essential oil on a packed column and on two types of open tubular column [12]. Instructional difficulties are sometimes experienced with those few chromatographers who prefer to deal with the packed column chromatogram and view the improved separation with misgivings. The qualitative and quantitative futility of that viewpoint can be emphasized by pointing out that the chromatogram can be made even simpler by omitting the stationary phase; all components will then emerge as a single peak.

On the other hand, there are occasions where the required degree of separation can be obtained on a packed column and separation on an optimized open tubular column results in more resolution than is required ("overkill") at the expense of longer analysis times. In situations of this type, some of the superior resolving power of the open tubular column can be traded off to yield equivalent (or even improved) separation in a fraction of the packed column analysis time, at higher sensitivities, and in a much more inert system (quantitative reliability is improved). As compared to packed columns and packed column analyses, the open tubular column can also confer distinct cost advantages [13].

The above points are illustrated in Fig. 1.2 [14], where the short capillary column delivers separation "equivalent" to that obtained in the much longer packed column analysis. Actually, the capillary resolution is superior; integrated values from the packed column analysis will include appreciable solvent; the

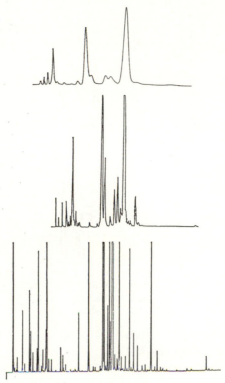

Fig. 1.1. Chromatograms of a peppermint oil on (top) a 6 ft × ¼ in. packed column, (center) a 500 ft × 0.03 in. stainless steel open tubular column, and (bottom) a 30 m × 250 μm fused silica open tubular (capillary) column. (Adapted from [12], p. 455, and reprinted by courtesy of Marcel Dekker, Inc.)

solute peaks are well removed from the solvent in the capillary analysis, and quantitation will be enhanced.

Returning to Fig. 1.1, the striking difference between the two sets of chromatographic results illustrated is best attributed to inequalities in the degree of randomness exhibited by the identical molecules of each individual solute. All of the identical molecules of each solute exhibit a narrow range of retention times in the bottom chromatogram, but as the ranges of retention times become greater, neighboring peaks exhibit overlap and resolution suffers. These behavioral differences between identical molecules can be attributed to three factors [10]:

1. The packed column offers solute molecules a multiplicity of flow paths, some short, the majority of average length, and some long. Hence identical molecules of each given solute would be expected to spend disparate times in the mobile phase. The open tubular column, on the other hand, has a single flow

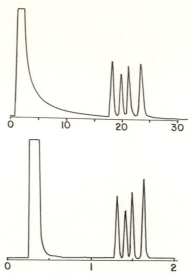

Fig. 1.2. Routine separations of methyl benzoates on (top) a packed column (analysis time 25 min) and (bottom) a 1.7 m × 0.25 mm glass open tubular column.

path, and molecules would be expected to exhibit mobile phase residence times that were much more nearly identical.

2. A similar rationale can be drawn for the randomness of stationary phase residence times. There is much more stationary phase in the packed column, and the film thickness is nonuniform. Thicker regions of stationary phase would be expected to occur in particle crevices and where two or more coated particles come into contact. A solute molecule dissolving in a thinner region of stationary phase would become dispersed and then reemerge to the moving gas phase in a relatively short period; an identical molecule, dissolving in a thicker patch of stationary phase, would take a longer time to reemerge; the times that identical molecules of a given solute spent in stationary phase would be quite diverse. In the open tubular column, the stationary phase is in a thinner and much more uniform film; the ranges of times that identical molecules spent in stationary phase would be expected to be much narrower.

3. It was previously mentioned that solute volatilities (i.e., solute vapor pressures) constitute an important variable in gas chromatography. The vapor pressure of a solute is an exponential function of the absolute temperature; a minor shift in temperature can have a major effect on vapor pressure. At our present state of instrumental development, solute temperatures are controlled via the column oven: the oven air conveys heat to the column wall, the column wall conducts heat to the particle of solid support in contact with it, this conducts heat to the stationary phase with which it is coated and to the next particle of solid

support, etc. Packed column support materials, however, are poor heat conductors. A temperature range must exist across any transverse section of the packed column; the range will be greater for larger-diameter columns and for higher program rates, but even in an isothermal mode a temperature range will exist across the column. Molecules whose flow path is in the center of that column will be at a lower temperature and exhibit lower vapor pressures and larger values of k (*vide infra*) than identical molecules whose flow paths are closer to the column wall. The fact that their flow paths undoubtedly switch back and forth does not compensate for this variation; it is one more factor causing identical molecules to exhibit a broadened range of retention times. In the fused silica column, the stationary phase exists as a thin film deposited directly on the inner wall of a tube of very low thermal mass; there should be no temperature variation across any transverse section of the column, provided that the column is heated only by convection. The latter point is an important distinction between the oven requirements for packed and capillary columns and is considered again in Chapter 7.

Our goal in chromatography can now be better defined. Gas chromatography should be performed under conditions where (1) solute molecules undergo many transitions (vaporizations and re-solutions), (2) identical molecules of each solute exhibit the narrowest possible range of retention times (i.e., the chromatographing band formed by each molecular species is short, hence the standard deviation of the resultant peak is small), and (3) the separation achieved is maintained to the highest degree possible.

1.4   Abbreviated Theory of the Chromatographic Process

The primary objective of this book is to introduce practical considerations involved in gas chromatography, including the selection, installation, evaluation, and use of open tubular glass capillary columns. Some knowledge of gas chromatographic theory is essential to the attainment of this goal, but this section is intended as neither a comprehensive nor a rigorous treatment of chromatographic theory. Theoretical considerations have been well covered elsewhere (e.g., [5,15–19]). In an attempt to avoid contributing further to the confusion caused by a variety of nonuniform "systems" of nomenclature, the symbols and nomenclature used throughout this discussion are based largely on those suggested by the International Union of Pure and Applied Chemistry (IUPAC) [20] and the American Society for Testing and Materials (ASTM) [21] and are detailed in the Appendix.

A compound subjected to the gas chromatographic process (a "solute") is, on injection into the column, immediately partitioned between the mobile phase and the stationary phase. Its apportionment between the two phases is reflected by the distribution constant K_D, defined as the ratio of the weights of solute in equal volumes of the stationary and mobile phases:

$$K_D = \frac{\text{concentration per unit volume stationary phase}}{\text{concentration per unit volume mobile phase}} = \frac{c_S}{c_M} \quad (1.1)$$

K_D is a true equilibrium constant, and its magnitude is governed only by the compound, by the stationary phase, and by the temperature. Polar solutes would be expected to dissolve in, disperse through, and form intermolecular attractions with polar phases to a much greater degree than would hydrocarbon solutes exposed to the same stationary phase. Logically, the K_D of a polar solute in a polar phase is higher than the K_D of the hydrocarbon of corresponding chain length in the same polar phase. As the temperature of the column is increased, both types of solute exhibit higher vapor pressures and their K_D values (c_S/c_M ratios) decrease, although (in this polar stationary phase) those of the polar solute remain larger than those of the hydrocarbon. Among the members of a homologous series, of course, higher homologs have lower vapor pressures and higher K_D values.

During its passage through the column, a solute spends a fractional part of its total transit time in the stationary liquid phase and the remainder in the mobile gas phase. The mobile phase residence time can be determined by direct measurement. Whenever a solute emerges to the mobile phase, it is transported toward the detector at the same rate as the mobile phase; hence everything must spend the same length of time in the mobile phase. This mobile phase residence time can therefore be determined by timing the elution of a solute that never enters the stationary phase, but spends all its time in the mobile phase; ideally, this could be determined by timing the transit period of an injection of mobile phase (carrier gas), but detection would be impractical. Methane is normally used for this purpose, and although it is now recognized that methane does have a discrete stationary phase residence time, it is assumed that this is minuscule and can be ignored in columns of "standard" stationary phase film thickness and at reasonable column temperatures.

The column residence time for methane is assigned the symbol t_M, and, as discussed earlier, a solute in the mobile phase is transported toward the detector at the same velocity as the mobile phase; hence everything spends time t_M in mobile phase. This is the "gas holdup volume" (or "gas holdup time") of the system. The total retention time is equal to the mobile phase residence time t_M plus the stationary phase residence time. It therefore follows that the stationary phase residence time or the "adjusted retention time" t_R' is

$$t_R - t_M = t_R' \quad (1.2)$$

Ideally, solute bands will be introduced into the column in such a way that they occupy a very short length of the column (see injection mechanisms in Chapter

3). It is highly desirable that the length of each band increase by a minimum amount as the solute bands traverse the column. Toward this goal, we require that the ranges of retention times exhibited by identical molecules be extremely narrow, i.e., that the standard deviation exhibited by each molecular species is small. As these tight, concentrated bands leave the column, they can be delivered to the detector as narrow, sharp peaks. In actuality, even if the times that identical molecules spend in mobile phase and in stationary phase are precisely the same, other factors such as longitudinal diffusion [occurring primarily in the mobile (gas) phase and to a very much lower degree in the stationary phase] cause lengthening of the solute bands during the chromatographic process. The centers of the bands of solutes that have different K_D values will become increasingly separated as they progress through the column, but if the range of retention times exhibited by identical molecules is large or if longitudinal diffusion is excessive, band lengthening may cause the trailing edge of the faster component to interdiffuse with the leading edge of the slower component, resulting in incomplete separation and overlapping peaks. Hence the efficiency with which two components can be separated is governed not only by their relative retentions (see below) but also by the degree of band lengthening that occurs. Insofar as the column is concerned, the separation efficiency is inversely related to the degree of band lengthening. All other things being equal, a minimum degree of band lengthening occurs per unit of column length in a column of high efficiency, and a higher degree of band lengthening occurs per unit of column length in a less efficient column. The term "band broadening" is often used in describing these phenomena; as discussed in later chapters, bandwidths are constant and limited by the column diameter; it is really the lengths of those bands that are of concern, because long bands lead to broad peaks.

Inasmuch as both are methods for separating mixtures of volatile compounds, it is not surprising that gas chromatography was promptly compared with the process of fractional distillation, and distillation terminology (i.e., "theoretical plates") was employed (albeit imperfectly; see below) to describe gas chromatographic separation efficiencies.

As detailed above, the separation efficiency of a gas chromatographic column is related to the degree to which a solute band lengthens (which correlates with peak width and affects the standard deviation of the peak, σ) relative to the time the band requires to traverse the column (its retention time t_R). The "number of theoretical plates" n is defined as

$$n = (t_R/\sigma)^2 \qquad (1.3)$$

where t_R is the time (or distance) from the point of injection to the peak maximum, and σ is the standard deviation of the peak. To avoid the necessity of determining σ, the peak is assumed to be Gaussian (which it probably is not; see below), and the problem is simplified. For a Gaussian peak, peak width at base

(w_b) is equal to 4.0σ, and peak width at half height (w_h) is equal to 2.354σ (Fig. 1.3). Substitution yields the relationships

$$n = (4t_R/w_b)^2 = 16(t_R/w_b)^2 \tag{1.4}$$

and

$$n = (2.35t_R/w_h)^2 = 5.54(t_R/w_h)^2 \tag{1.5}$$

Because the w_h peak width measurement can be made directly and with greater precision, most workers prefer Eq. (1.5). The same units must, of course, be used for the t_R and w measurements. As mentioned above, these equations assume Gaussian-shaped peaks, and it is very doubtful that the average chromatographic peak is indeed Gaussian. In most cases the deviation is not great, but theoretical plate measurements cannot be applied with any real meaning to peaks that are obviously malformed or asymmetric. Peak asymmetry may be caused by (among other things) reversible adsorption or overloading. These phenomena are discussed in later sections.

Obviously, longitudinal diffusion of components in the column is one factor

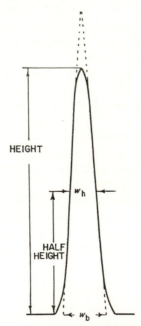

Fig. 1.3. Characteristics of a Gaussian peak. The peak width at the base (which must be determined by extrapolation) is equal to 4 standard deviations (σ), and the peak width at half-height (which can be measured directly on an on-scale peak) is equal to 2.35 standard deviations.

affecting band lengthening; lower- molecular-weight solutes (usually characterized by smaller K_D values) would be expected to diffuse to a greater degree per unit time than higher-molecular-weight solutes. Stationary phase diffusivity also affects values of both t_R and w_h (Chapter 5). Consequently, the value of n is a function not only of the column and of the solute, but also of the temperature, the type of carrier gas, and the degree to which that gas is compressed (i.e., the pressure drop through the column). These relationships are discussed in later chapters.

The determination of the gas holdup volume t_M by methane injection was discussed briefly above. It is also possible to estimate the value of t_M by calculation [1]. Although t_M is an integral part of t_R and makes a positive contribution to the theoretical plate number, it obviously contributes nothing to the separation process; indeed, simply by insertion of a long empty fine-bore tube between the injection assembly and the front end of the column, one could achieve very large values of t_M, leading to still larger values for t_R. This would give a grossly inflated value for the number of theoretical plates, n, of the column. This possibility can be eliminated by dealing with an adjusted retention time t_R':

$$t_R' = t_R - t_M \qquad (1.6)$$

This value is used in calculating the number of usable or effective theoretical plates, N:

$$N = 5.54[(t_R - t_M)/w_h]^2 = 5.54(t_R'/t_M)^2 \qquad (1.7)$$

Longer columns (of identical efficiency per unit length) will possess more theoretical plates, although because of complicating factors such as the increased pressure drop, the relationship approximates linearity only at the optimum mobile phase velocity ($\bar{u}_{opt}$; see below).

Efficiencies are sometimes expressed as the number of theoretical plates per meter of column length, e.g., n/m or N/m. More often, however, the inverse value—the length of column occupied by one theoretical plate—is used. Once again, distillation terminology is employed, and this is termed the "height equivalent to a theoretical plate." It is usually expressed in millimeters and is given the symbol h:

$$h = L/n \qquad (1.8)$$

where L is the column length. Similarly, the height equivalent to one effective theoretical plate is given the symbol H:

$$H = L/N \qquad (1.9)$$

Obviously, smaller values of h (or H) indicate higher column efficiencies and greater separation potentials. The term h_{min} (or H_{min}) is used to express the value of h (or H) when the column is operating under optimum flow conditions.

Inasmuch as the values of n and N are affected by the column temperature, the test compound, and the nature of the carrier gas, these parameters must also affect h and H.

As previously described, a solute undergoing separation spends a certain proportion of its transit time in the mobile (gas) phase and the remainder in the stationary (liquid) phase. The sum of these times is, of course, its observed retention time t_R. During the periods when a substance is in the mobile phase, it is moving toward the detector at the same velocity as the carrier gas. Therefore, regardless of their (total) retention times, all substances spend the same length of time, equal to t_M, in the mobile phase. The time spent in the stationary phase will therefore be equivalent to the adjusted retention time t_R'. The partition ratio* k is defined as the amount (not concentration) of a solute in stationary phase compared to the amount of that solute in mobile phase. This is proportional to the time the solute spends in stationary phase relative to the time it spends in mobile phase:

$$k = t_R'/t_M \tag{1.10}$$

Theoretical plate numbers (n) are, of course, always larger than effective theoretical plate numbers (N), and for a given set of conditions the magnitude of the difference is a function of the partition ratio (k):

$$N = [k/(k + 1)]n \tag{1.11}$$

and

$$n = [(k+1)/k]N \tag{1.12}$$

For a solute whose partition ratio is very large (i.e., large k and long retention), t_M is such a minuscule portion of t_R that (for all practical purposes) $t_R = t_R'$. Hence with very large-k solutes, $k = (k + 1)$, and $n = N$.

Figure 1.4 illustrates the effect that the partition ratio of the test compound has on the calculated maximum theoretical values of n and N for a 50 m $\times$ 0.25 mm column. Experimental determinations performed on a series of several solutes rarely exhibit the smooth relationships shown in Fig. 1.4. Part of the reason for this discrepancy lies in the fact that the data for Fig. 1.4 are calculated at the optimum mobile phase velocity for each individual solute, which is a function of the solute partition ratio. In normal experimental determinations, the mobile phase velocity is constant during the separation of the entire series of solutes. Hence some solutes are chromatographed above (and others below) their optimum velocities; this will have adverse effects on the theoretical plate numbers generated, and the magnitude of those effects will be different for each solute.

---

*The term "partition ratio" is preferred by IUPAC [20], while "capacity factor" is preferred by ASTM [21]; see the Appendix.

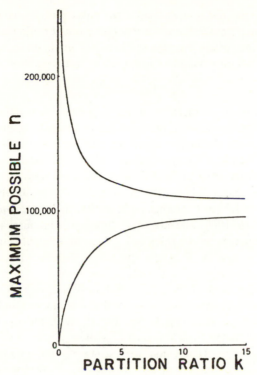

Fig. 1.4. Maximum possible plate numbers for a 50 m × 0.25 mm column. Values calculated at $\bar{u}_{opt}$ for each value of k. Upper curve, theoretical plates n; lower curve, effective theoretical plates N.

Logically, the proportion of the analysis time that a substance spends in the stationary phase relative to the time spent in mobile phase, k, must be related to its distribution coefficient K_D. This relationship hinges on the relative availability of (i.e., the volumes of the column occupied by) the mobile (gas) and stationary (liquid) phases (i.e., the column "phase ratio") and is given the symbol β:

$$\beta = V_M/V_S \tag{1.13}$$

Referring back to Eq. (1.1), it can be seen that the distribution constant K_D can also be defined as

$$K_D = \frac{\text{amount in stationary phase}/\text{volume of stationary phase}}{\text{amount in mobile phase }/\text{ volume of mobile phase}}$$

or

$$K_D = \frac{\text{amount in stationary phase}}{\text{amount in mobile phase}} \times \frac{\text{volume of mobile phase}}{\text{volume of stationary phase}}$$

The latter fraction has just been defined as the phase ratio β, and the former is the same as the partition ratio k. From this emerges a relationship that will be utilized later in rationalizing certain injection mechanisms and in understanding the inter-relationships of several parameters in retention and separation characteristics:

$$K_D = \beta k \qquad (1.14)$$

It is apparent that β is a measure of the "openness of the column," and one would expect the phase ratios of open tubular columns to be appreciably larger than those of packed columns, in which the packing not only limits the volume available for mobile phase but also increases the support area over which thicker films of the stationary phase are distributed. The phase ratios of packed columns are usually in the range of about 5 to 35, whereas in open tubular columns with "normal" (i.e., <0.5 μm) stationary phase film thickness, the values usually fall in the range 50 to 1000. The advent of "bonded" stationary phases has led to open tubular columns with ultrathick (e.g., 3, 5, and 8 μm) films, whose phase ratios as may be as low as 10. These special-purpose columns are discussed in later chapters.

The inner surface area of the open tubular column (which at constant film thickness governs V_S) varies directly with column diameter, while the volume of the column (which governs V_M) varies directly with the square of the inner radius, i.e., the distance from the center of the column to the surface of the stationary phase coating. Hence both the diameter of the column and the thickness of the stationary phase film exercise effects on the phase ratio of open tubular columns. For the more commonly used columns, d_f values range from 0.1 to 1.0 μm, and the column diameter is of the order of 200 to 530 μm; hence the effect of d_f on the mobile phase volume of the column has usually been neglected and the phase ratio expressed as

$$\pi r^2 / 2r\pi d_f = r/2d_f \qquad (1.15)$$

Inasmuch as the volume of stationary phase is $2r\pi d_f$ per unit of column length, the mobile phase volume is more precisely

      total (wall-to-wall) volume $-$ volume occupied by stationary phase

or $\pi r^2 - 2r\pi d_f$ per unit of column length. A more accurate measure of the phase ratio is therefore

$$\beta = (\pi r^2 - 2\pi r d_f)/2\pi r d_f = (r - 2d_f)/2d_f \qquad (1.16)$$

Because $[(r - 2d_f)/2d_f] + 1 = r/2d_f$, the values calculated from Eq. (1.15) are exaggerated by one unit; except for columns of very low phase ratio (e.g., thick film or small diameter), differences between values calculated from Eqs. (1.15) and (1.16) insignificant. Of even greater import is the fact that because column production tolerances usually approximate ± 0.5 μm for the radius of the tubing and $\pm 5\%$ for d_f, the difference between actual and calculated phase ratios usually exceeds the discrepancy between results derived from the different equations.

The phase ratio of a column can be determined by measuring the k of a solute of known K_D, but the accuracy may be influenced by slight variations in the chemical composition of the stationary phase (as compared to the column used in the initial determination of K_D) or by temperature deviations, both of which would affect K_D. Distribution constants for many of the n-paraffin hydrocarbons on several stationary phases at specific temperatures have been published (e.g., [22–24]); some of those values are reproduced in Table 5.3. Alternatively, the phase ratio can be determined by comparison with another column of known phase ratio. If it is assumed that the known quantity of stationary phase is uniformly distributed on the inner periphery of open tubular columns coated by a static technique, the phase ratio for that column can be calculated from Eq. (1.15) or (1.16). A test compound can then be chromatographed, and after measuring t_M, k for that compound can be calculated [Eq. (1.9)]. From Eq. (1.14), K_D for the test compound can then be determined, and, as detailed earlier, this value will be the same for that solute on any column containing the same stationary phase at the same temperature. When the test compound is then chromatographed on the new column under the same conditions, $K_{D(1)} = K_{D(2)}$, and β for the second column can readily be calculated from the relationship

$$\beta_{(2)} = \beta_{(1)}(k_{(1)}/k_{(2)}) \tag{1.17}$$

1.5   Separation of Components

The degree to which two components are separated is a function of (1) the ratio of their retention times and (2) the sharpness of the peaks (n). The ratio of the adjusted retention times of two components, 1 and 2, is termed their relative retention:

$$\alpha = t'_{R(2)}/t'_{R(1)} \tag{1.18}$$

From Eq. (1.9) it is readily apparent that the relative retention can also be expressed as

$$\alpha = k_{(2)}/k_{(1)} = K_{D(2)}/K_{D(1)} \tag{1.19}$$

By convention, α is never less than 1.0, so the function of the second (or more retained) solute is always used as the numerator. Solute pairs with large α values can be separated relatively easily even on low-resolution columns, but as this ratio approaches unity, columns with increasingly larger numbers of theoretical plates are required to achieve separation. Alternatively, of course, one can sometimes select another stationary phase in which the relative retention of those components is larger.

The degree of separation of two components (1 and 2) is termed resolution R_s:

$$R_s = 2(t_{R(2)} - t_{R(1)})/(w_{b(1)} + w_{b(2)}) \qquad (1.20)$$

This definition, however, requires extrapolation to determine the widths of those peaks at their bases. If peaks are assumed to be Gaussian, then because of the relationship between w_b and w_h discussed earlier in connection with Eq. (1.4) and shown in Fig. 1.3, resolution can also be expressed as

$$R_s = 1.18(t_{R(2)} - t_{R(1)})/(w_{h(1)} + w_{h(2)}) \qquad (1.21)$$

Hence resolution (or component separation) is related to the degree of peak broadening and retention time differences. As shown in Fig. 1.5, a resolution of 1.0, while separating idealized peaks, actually results in a considerable degree of overlap. A resolution of 1.5 will usually achieve baseline separation, but asymmetry or tailing can cause complications.

From a knowledge of the relative retention of two compounds, one can closely

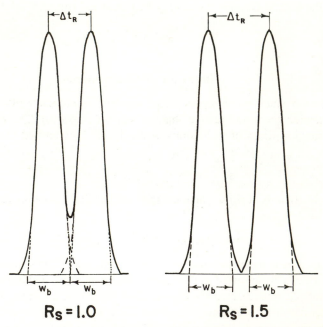

Fig. 1.5. Resolution and component separation. A resolution of 1.5 is usually equivalent to "baseline separation" with symmetrical peaks.

approximate, under that specific set of conditions, the number of effective theoretical plates required for any desired degree of resolution:

$$n_{req} = 16R_s^2 \, [\alpha/(\alpha - 1)]^2 \, [(k + 1)/k]^2 \qquad (1.22)$$

where k is the partition ratio of the second component. This can be rearranged to
yield

$$R_s = \tfrac{1}{4}\sqrt{n}[(\alpha - 1)/\alpha] \, [(k/k + 1)] \qquad (1.23)$$

Several practical conclusions can be drawn from Eqs. (1.22) and (1.23). First,
resolution is proportional to the square root of the number of theoretical plates (or,
at $\bar{u}_{opt}$, to the column length); i.e., in order to double the number of theoretical
plates delivered by a 25-m column operating at its optimum carrier gas velocity, a
100-m column of equivalent quality would be required, and the analysis would
seem to take four times as long. In actuality, because optimum gas velocities vary
inversely with column length (Chapter 5), the latter analysis would take more than
four times as long, if other conditions remained unchanged.

Second, resolution is influenced by both the relative retention α and the
partition ratio k. This interrelationship is partly responsible for the bewildering
array of stationary phases available to the packed column chromatographer. Even
at its best, the packed column is capable of delivering only a few thousand
theoretical plates; the packed column chromatographer therefore frequently requires stationary phases of greater selectivity—i.e., stationary phases that generate larger α values for those particular solutes under those specific conditions.
The open tubular column can achieve much larger theoretical plate numbers. In
many cases, a stationary phase unsatisfactory for packed column use (because at
that restricted number of theoretical plates, the relative retentions of the solutes
were too small) may deliver complete separation, shorter analysis times, and
higher sensitivity in the open tubular system. These interrelationships may be
complicated by the fact that because the phase ratio of the open tubular column is
larger, solute partition ratios are necessarily smaller [Eq. (1.11)]. Where a very
small partition ratio is increased e.g., from 0.01 to 1.0, the factor $[(k + 1)/k]^2$
drops from 10,000 to 4 and becomes a less significant multiplier (see Table 5.1).
One of the column parameters to be discussed in a later chapter is that of
stationary phase film thickness. Thin-film columns are very useful for the analysis of larger and high-boiling solutes but are rarely satisfactory for general-
purpose use. Theoretical plate numbers vary indirectly* and partition ratios vary
directly with the thickness of the stationary phase film; under the same conditions, columns with thinner films exhibit higher theoretical plate numbers, but

---

*This generality is limited to high-diffusivity phases and d_f values in excess of about 0.4 μm
(Chapter 5).

because solutes have smaller partition ratios in those columns, separation may be far from satisfactory in spite of the larger plate numbers. As the partition ratio of the test compound increases, the factor $[(k + 1)/k]^2$ becomes less and less significant. These relationships are discussed in detail later.

1.6   Effect of Carrier Gas Velocity

A number of valuable concepts are embraced by the van Deemter equation [25], which examines the efficiency of the column (or, more accurately, of the entire system; see later chapters) as a function of the rate of mobile phase (carrier gas) flow. Gas flows through the open tubular column are very low when expressed in normal volumetric terms; it is also worth noting that it is the linear velocity of the mobile phase rather than its volumetric flow that conducts solutes through the column. A few workers still report gas flows in cubic centimeters per minute (or in milliliters per minute) (F), but most prefer to speak of an "average linear carrier gas velocity" in centimeters per second ($\bar{u}$). It can be shown that

$$F = 0.6\pi r^2 \bar{u} \qquad (1.24)$$

hence

$$\bar{u} = 1.67F/\pi r^2 \qquad (1.25)$$

where F is the volumetric flow in cubic centimeters per minute, r the radius of the capillary in millimeters, and $\bar{u}$ the average linear carrier gas velocity in centimeters per second. Usually F is measured with a small-volume bubble flowmeter and u is determined by methane injection:

$$\bar{u} = L/t_M \qquad (1.26)$$

Column efficiencies are expressed in terms of h, and of course small values of h denote large values of n and greater separation potentials. The width of the peak is a factor in the determination of n, and as a solute reaches the end of the column and is delivered to the detector, the width of the peak generated by the detector is a function of the time required for the carrier gas to conduct that solute from its position at the end of the column to the detector. This is partly a function of the length of the solute band in the column; it is also a function of the distribution constant of the solute, as discussed earlier. Because the distribution constant is an exponential function of the column temperature, peak widths are affected by the temperature of the column at the time of their elution to the detector. Obviously, this has an effect on column efficiency measurements based on peak widths, and it complicates efficiency determinations that are conducted under conditions of temperature programming. In most cases, comparative evaluations should employ isothermal conditions.

In its abbreviated form, the van Deemter equation can be written:

$$h = A + B/\bar{u} + C\bar{u} \tag{1.27}$$

where A includes packing and multi-flow-path factors, B is the longitudinal diffusion term, C the resistance to mass transfer from the mobile phase to the stationary phase and from the stationary phase to the mobile phase, and $\bar{u}$ the average linear velocity of the mobile gas phase. Open tubular columns contain no packing and the A term becomes zero, reducing the van Deemter equation to a form known as the Golay equation [7]:

$$h = B/\bar{u} + C\bar{u} \tag{1.28}$$

To achieve the maximum separation potential, the goal is obviously the lowest possible value for h (or H), which is equivalent to the highest possible value of n (or N). Elimination of the A term from the van Deemter equation (the Golay equation) is one step in this direction. As h varies indirectly with the value of $\bar{u}$ in the B term and directly with the value of $\bar{u}$ in the C term, there must exist some optimum value of $\bar{u}$ at which any given system will achieve the highest efficiency for a solute of a given partition ratio. This can be calculated or determined graphically, and the interested reader is referred to more general or theoretical references (e.g., [16–18]). It should be mentioned at this point that the skilled chromatographer usually operates at this optimum velocity only to prove the system and performs most analyses at higher gas velocities, generating results that are commensurate with the degree of separation required, the efficiency of the column, the range of partition ratios embraced by the sample solutes, and the time available for the analysis (see later chapters).

Because plate numbers are a function of the square of the t_R/w_h ratio, h will be decreased by any factors that make this ratio larger and increased by any factors that make it smaller. Although w_h is affected by the length of the starting band (i.e., injection efficiency) and the distribution constant (discussed above), if these are held constant w_h reflects lengthening of the band occasioned by its passage through the column (i.e., column efficiency) relative to its transit time (which increases as the solute spends more time in the stationary phase or experiences a greater number of interphase transitions). One cause of band lengthening is longitudinal diffusion. As discussed in Chapter 5, diffusion constants are of the order of $0.15–0.5$ cm²/sec in mobile phase and approximately 10^{-6} cm²/sec in stationary phases of higher diffusivity. The contribution of diffusion occurring in the stationary phase is extremely small in comparison with mobile phase diffusion, and for most practical purposes stationary phase diffusion can be ignored with respect to band lengthening in open tubular columns. Diffusivity of solute molecules in the gas phase varies inversely with both the density of the carrier gas (i.e., diffusion is greater in hydrogen than in helium than in nitrogen) and the size of the solute molecule (under the same conditions, hexane diffuses more rapidly in helium than does decane). Temperature also has an effect on diffusivity; as the temperature increases, the greater kinetic energy of the individual solute molecules encourages diffusion, but the increased viscosity of the mobile phase discourages diffusion. The gas phase concentration of the solute also plays a role in diffusion: within a given gas under a given set of conditions, a more concentrated vapor plug diffuses more rapidly than a less concentrated plug of the same vapor.

The two extremes under gas chromatographic conditions would be a stop-flow situation, where band lengthening occasioned by diffusion would be maximum, and a gas velocity sufficiently high to carry the solute band through the column before gas phase diffusion can make a measurable contribution to band lengthening. These considerations are, of course, oversimplifications that assume laminar flow and ignore concentration effects.

The retention time of a solute is governed in part by the time it spends in the stationary phase, i.e., how effectively it utilizes the stationary phase on its route through the column. As an individual solute molecule vaporizes from the stationary phase, it diffuses laterally through the mobile phase (carrier gas) to a new solution site in the stationary phase; during this lateral diffusion it is at the same time being conducted through the column (and past available stationary phase solution sites) by the carrier gas stream. As low carrier gas velocities, a given solute molecule undergoes more re-solutions into the stationary phase (hence more vaporizations from the stationary phase) than it does at high velocities. These re-solution and vaporization processes have been termed "resistance to mass transfer." "Mass transfer" usually implies that special contributions on the part of those molecules at the interphase (or interfacial boundary) have been taken into account. Such "surface tension phenomena" probably do play at least minor roles in gas chromatography, but, as currently used, the van Deemter equation treats solute diffusion in the mobile (and in the stationary) phase as the defined barrier to this step, and "mass transport" seems a more appropriate term.

It should now be apparent that the best results would be achieved under conditions of infinitely slow longitudinal diffusion and infinitely fast lateral diffusion. Obviously, this will not be possible.

A primary difference between packed and open tubular columns is that the former have thicker and more irregular films of stationary phase, separated by smaller gas volumes. The difference becomes readily apparent if the Golay equation is written so the two contributions to the "resistance to mass transport" term can be differentiated, i.e., expressed as "mass transport from mobile phase to stationary phase" and "mass transport from stationary phase to mobile phase":

$$h = B/\bar{u} + (C_M\bar{u} + C_S\bar{u}) \qquad (1.29)$$

With packed columns the C_M term is vanishingly small in comparison to the C_S term; in open tubular columns the reverse is true, and the C_S term can be ignored. (This generality is true for "normal" open tubular columns; for columns having thicker films of stationary phase and/or stationary phases of low diffusivities, C_S becomes increasingly important; see below.) The significance of this distinction and how it affects the choice of stationary phase and carrier gas are considered in Chapter 5.

Again, it can be helpful to (incorrectly) visualize gas chromatography as a stepwise process. As the linear velocity of the carrier gas approaches zero, a vaporizing solute molecule is conducted (longitudinally) through a minimum length of column while it diffuses laterally across the column to its next stationary phase solution site. Under these conditions of very low mobile phase velocity, each solute molecule would experience a maximum number of solution and vaporization steps during its passage through the column, and while retention times would become very large, the value of h—as influenced only by this parameter and ignoring band lengthening occasioned by longitudinal diffusion— would approach zero and be dictated largely by the efficiency of the injection process (Chapter 3). As the velocity of the mobile phase is increased, the length of column through which the vaporizing solute molecule is swept before it can diffuse across the gas volume and dissolve in a new solubility site also increases. Assuming that the rate of transverse diffusion remained constant, the degree to which an average solute molecule is denied access to stationary phase would be directly proportional to the mobile phase velocity. High carrier gas velocities decrease both the time spent in the stationary phase (fewer solution steps) and the time spent in the mobile phase. Unless the velocity becomes so high that the flow becomes turbulent or solute equilibrations are affected, solute partition ratios should remain the same.

In Fig. 1.6, the line representing the resistance to mass transport (lateral diffusion) intersects the ordinate slightly above the abscissa as detailed above. The slope of the line varies directly with column length, column diameter (open tubular columns), and density of the carrier gas; it usually varies inversely with column temperature. In other words, anything that inhibits the process of lateral diffusion—lower column temperature (usually, but the resultant decrease in gas viscosity may complicate this parameter), higher-molecular-weight (i.e., higher-k) solutes, denser carrier gases (longer columns require larger pressure drops, resulting in a denser carrier gas)—or increases the distance over which that diffusion must occur (larger-diameter columns) will produce a steeper slope.

But the negative contribution of longitudinal diffusion to separation must also be considered. Component resolution is enhanced by preserving to the largest degree possible the separation achieved by the multiple vaporization steps considered above. A major detriment is longitudinal diffusion leading to remixing of partially resolved solute bands. Longitudinal diffusion occurs almost entirely in

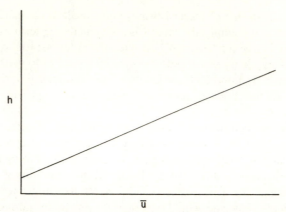

h

ū

Fig. 1.6. Effect of average linear carrier gas velocity ū on column efficiency h as determined only by the "resistance to mass transport" (C) term of the van Deemter equation. The longitudinal diffusion (B) term has not been taken into account. At very low carrier gas velocities, solute molecules encounter more stationary phase per unit length of column; they undergo a large number of vaporizations (and re-solutions); n is large, h is small. At higher carrier gas velocities, solute molecules encounter less stationary phase per unit length of column and can undergo fewer vaporizations (and re-solutions); insofar as this factor only is concerned, n varies indirectly (and h varies directly) with the average linear carrier gas velocity.

the mobile gas phase; the band-lengthening contribution of diffusion in the stationary phase is normally minuscule. Diffusion is a function of time, and the time the solute bands spend in the mobile gas phase is a function of the mobile phase velocity. At a very low mobile phase velocity, the negative contribution of longitudinal diffusion is large, n decreases, and h becomes large; at a high

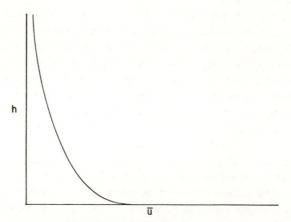

h

ū

Fig. 1.7. Effect of average linear carrier gas velocity ū on column efficiency h as determined only by the longitudinal diffusion (B) term of the van Deemter equation. This does not take into account the "resistance to mass transport" (C) term (Fig. 1.6). See text for discussion.

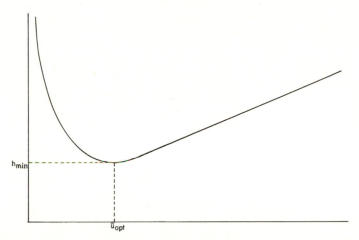

Fig. 1.8. Total effect of average linear carrier gas velocity $\bar{u}$ on column efficiency h. The van Deemter curve represents the summation of the curves in Figs. 1.6 and 1.7.

mobile phase velocity, the negative contribution of longitudinal diffusion is small, n remains large, and h is small. Because diffusion is also a function of concentration, the net result of longitudinal diffusion as influenced by the velocity of the mobile phase (ignoring mass transport) is a curve, as shown in Fig. 1.7.

The van Deemter curve [25] represents the sum of the longitudinal diffusion term ($h = B/\bar{u}$; Fig. 1.7) and the resistance to mass transport term ($h = C_M\bar{u} + C_S\bar{u}$; Fig. 1.6), as shown in Fig. 1.8. Experimental curves (see later chapters) vary from the usual theoretical curves in that they continue to curve upward at higher velocities rather than exhibit a straight line whose slope is dictated by the C term. This discrepancy is related to the fact that the average linear velocity of the mobile gas phase ($\bar{u}$) must be varied by changing the pressure drop through the column. The pressure drop, of course, affects the average density of the gas, which in turn exerts an influence on diffusivity and the slopes of both the B and C terms. Various correction factors have been introduced to help correct these discrepancies. The van Deemter curve and its use in evaluating variables in the chromatographic process are considered in greater detail in Chapter 5.

References

1. M. Tswett, *Ber. Dtsch. Bot. Ges.* **24,** 316 (1906).
2. M. Tswett, *Ber. Dtsch. Bot. Ges.* **24,** 384 (1906).
3. A. T. James and A. J. P. Martin, *Biochem. J.* **50,** 697 (1952).
4. N. H. Ray, *J. Appl. Chem.* **4,** 21 (1954).
5. Various authors, *in* ''Chromatography'' (E. Heftmann, ed.). Van Nostrand-Reinhold, Princeton, New Jersey, 1961.

6. Various authors, *in* "75 Years of Chromatography—A Historical Dialogue (L. S. Ettre and A. Zlatkis, eds.). Elsevier, Amsterdam, 1979.

7. M. J. E. Golay, *in* "Gas Chromatography 1957" (East Lansing Symposium) (V. J. Coates, H. J. Noebels, and I. S. Fagerson, eds.), pp. 1–13. Academic Press, New York, 1958; see also "Gas Chromatography 1958" (Amsterdam Symposium) (D. H. Desty, ed.), pp. 139–143. Butterworth, London, 1958.

8. D. H. Desty, J. N. Haresnip, and B. H. F. Whyman, *Anal. Chem.* **32,** 302 (1960).

9. R. Daneneau and E. Zerrener, *J. High Res. Chromatogr.* **2,** 351 (1979).

10. W. Jennings, *in* "Glass Capillary Gas Chromatography in Clinical Analysis" (H. Jaeger, ed.), Chapter 1. Dekker, New York, 1985.

11. W. Jennings, "Gas Chromatography with Glass Capillary Columns," 2nd ed. Academic Press, New York, 1980.

12. T. Shibamoto, *in* "Application of Glass Capillary Gas Chromatography" (W. Jennings, ed.), p. 455. Dekker, New York, 1981.

13. W. Jennings, "Comparisons of Fused Silica and Other Glass Columns for Gas Chromatography." Huethig, Heidelberg, 1981.

14. K. Yabumoto, personal communication (1982).

15. J. C. Giddings, "Dynamics of Chromatography," Part I. Dekker, New York, 1965.

16. A. B. Littlewood, "Gas Chromatography: Principles, Techniques, and Applications," 2nd ed. Academic Press, New York, 1970.

17. B. L. Karger, L. R. Snyder, and C. Horváth, "An Introduction to Separation Science." Wiley, New York, 1973.

18. C. Horváth and W. R. Melander, *in* "Chromatography" (E. Heftmann, ed.), Part A. Elsevier, Amsterdam, 1983.

19. M. L. Lee, F. Yang, and K. D. Bartle, "Open Tubular Column Gas Chromatography: Theory and Practice." Wiley (Interscience), New York, 1984.

20. International Union of Pure and Applied Chemistry, Commission on Analytical Nomenclature, Recommendations on Nomenclature for Chromatography, *Pure Appl. Chem.* **37,** 445 (1974).

21. American Society for Testing and Materials (ASTM), Committee E-19, "Standard Recommended Practice for Gas Chromatography Terms and Relationships," ASTM E 355-77. ASTM, Philadelphia, Pennsylvania, 1983.

22. L. Butler and S. J. Hawkes, *J. Chromatogr. Sci.* **10,** 518 (1972).

23. J. M. Kong and S. J. Hawkes, *J. Chromatogr. Sci.* **14,** 279 (1976).

24. W. Millen and S. J. Hawkes, *J. Chromatogr. Sci.* **15,** 148 (1977).

25. J. J. van Deemter, F. J. Zuiderweg, and A. Klinkenberg, *Chem. Sci.* p. 5 (1956).

CHAPTER 2

THE OPEN TUBULAR COLUMN

2.1 General Considerations

The comparisons of packed and open tubular columns considered in Chapter 1 make quite obvious some of the advantages of the latter: in contrast to packed columns, open tubular columns can yield improved separations, shorter analysis times, and higher sensitivities and make it possible to construct much more inert analytical systems. In later chapters, this list of advantages will be expanded to include improved quantitation and less expensive installation and operational costs [1,2]. Although most knowledgeable chromatographers are well aware of the advantages offered by the open tubular column, most chromatographic efforts to this point in time continue to employ packed columns. Among the factors contributing to this are the following:

1. Initially, open tubular chromatographers had little choice but to construct their own columns. Until the mid-1970s, patents on the open tubular column were rigidly enforced and the commercially available columns were of inferior quality. Investigators who recognized the advantages to be gained and who wished to employ higher-quality columns were forced to fabricate their own. Column manufacture is an extremely labor-intensive process, and few users could justify activity in this area; most industrial (and many nonindustrial) users continued to employ packed columns.

2. Instrument- and operator-related problems were more common with open tubular columns. Because the pressure drop per unit length is lower, the open tubular column can have smaller-diameter tubing. Although such columns yield

higher efficiencies, they also place additional demands on the instrument and on the operator. Partially to circumvent those demands, attention was also soon directed to open tubular columns as large as 0.75 mm i.d. These very large open tubular columns tolerated relatively large carrier gas flows, and no special instrumental modifications were required. However, these columns were usually characterized by poor efficiencies and excessive activity. Nevertheless, they were widely used in certain industries. Segments of the petroleum industry that were less concerned about activity problems used smaller-diameter metal capillaries. Desty's machine [3] finally made long lengths of glass capillary tubing readily available and added great impetus to the development of the open tubular column.

3. Suitable instruments were not readily available. To retain a competitive position and remain commercially viable, instrument manufacturers design their products with a view toward accommodating the largest number of users, and commercial chromatographs were designed for packed columns. Investigators who fabricated capillaries had to learn to adapt packed column instruments to their special requirements. As the proportion of the market employing capillary columns increased, manufacturers began offering options which conferred capillary capability on the packed column instrument; many such conversions were often relatively crude and not very satisfactory. At this stage of development, the state of the art in instrumental design and instrumental modification resided not with the instrument manufacturer but with individual researchers.

4. The separation of some solutes could not be accomplished on the open tubular columns of that time but could be achieved on packed columns. The separation of highly volatile or very low molecular weight solutes (i.e., low-k solutes such as methane, ethylene, ethane) poses a formidable challenge to "normal" open tubular columns. This is partly due to the fact that the phase ratios of open tubular columns are larger ($\beta = V_M/V_S$); from Eq. (1.14), the partition ratio of a solute, k, varies inversely with β and must therefore be larger in packed columns, if all other factors are held constant. This is also related to the fact that t_M is smaller in packed than in capillary columns, with the result that solute partition ratios

$$\text{stationary phase residence times/mobile phase residence times}$$

are much larger in packed than in open tubular columns. The much larger partition ratios of packed columns could result in reducing n_{req} [Eqs. (1.22) and (1.23); Table 4.1] of extremely low-k solutes to a degree that more than compensated for the lower theoretical plate number of the packed column.

Much of the advantage that the packed column had for the separation of low-k (or highly volatile) solutes has now been eroded by the advent of low-β open tubular columns, i.e., open tubular columns with "superthick" films of crosslinked, surface-bonded stationary phase. Perhaps a more critical distinction is that an adsorptive packing could be used in packed columns designed for the analysis of fixed gases or very low molecular weight solutes; i.e., separations were based on gas–solid rather than gas–liquid chromatography. This distinction has also been achieved with (PLOT) porous layer open tubular and support-coated open tubular (SCOT) columns; their modern equivalents in fused silica columns are discussed later.

5. High-quality open tubular columns are readily available today, but they do place increased demands on the instrument and on the operator. In times past, chromatographers referred to the column as "the heart of the chromatographic system"; the overall performance of the system and the quality of the results obtained were most commonly dictated by the quality of the column. Column limitations could usually be attributed to the restricted powers of separation, which were usually related (1) to the limited theoretical plate numbers of the column and (2) to its catalytic and adsorptive properties. As a result, both inadequacies in instrument design and imperfections in operator technique were well masked. Columns available today can reduce to the vanishing point those restrictions previously imposed by the column, and instrumental design and/or operator technique become limiting; analysts (and instrument manufacturers) accept this challenge with varying degrees of enthusiasm.

6. From the above considerations, it is understandable that open tubular chromatography was reserved to those willing (and able) to devote significant and seemingly unrelated investments of labor, time, and instrumental modification to their analtyical endeavors. Most industrial scientists are not in such a position; they are instead under pressure to produce results, and their analytical endeavors must frequently employ equipment that can be purchased and operated *per se*.

Certainly these factors influenced those engaged in analytical methods development; the applicability of a method would be significantly reduced if it required (1) modification of an instrument and (2) construction of a suitable column. As one example, most of the gas chromatographic procedures developed by the Environmental Protection Agency are based on packed column analyses and required significant investments in the form of grants and contracts. There is a reluctance to acknowledge that most of those procedures are already seriously outdated; only recently, and as a result of overwhelming evidence (much of which is generated by the more progressive members of that agency), have some procedures been amended to include open tubular chromatography. Other agencies have circulated internal memoranda specifying that open tubular columns will be employed in any new gas chromatographic methods.

Partly because the academician has a tendency to teach from a historical perspective and partly because of the restrictions considered above, a large proportion of gas chromatographic instruction, both academic and industrially oriented, continues to be based on the packed column. This, in turn, leaves many industrial and academic scientists in a position where they feel unfamiliar and uncomfortable with open tubular chromatography; the preponderance of our analytical procedures have been designed around packed columns. Although procedures are usually readily transferable and the challenge that the open tubular column poses to the competent packed-column chromatographer is short-lived, there has been considerable resistance to abandoning the packed column. The availability of wide-diameter open tubular columns of flexible fused silica, with state-of-the-art deactivation and a wide range of bonded stationary phases, has overcome much of this inertia; an ever-increasing proportion of our gas chromatographic analyses are being converted to open tubular columns.

2.2   The Tubing

A major development that enabled many scientists to enter the field of open tubular gas chromatography occurred in 1960, when Desty *et al*. [3] suggested an elegantly simple machine capable of drawing and coiling long lengths of capillary tubing from conventional (i.e., soda lime, borosilicate) glasses. Operating temperatures varied with the type of glass being handled but were approximately 600°C for both the softening oven and the bending tube; capillary tubing was produced at a rate of approximately 1–2 cm/min. Heavier-walled tubing was more general (~0.25 mm i.d. and 0.75 mm o.d.), but the advantage of increased flexibility conferred by thinner walls (~0.25 mm i.d., 0.5 mm o.d.) was noted (e.g., [4]).

This machine made it feasible for many investigators to produce glass capillary tubing and, from that tubing, columns. Material costs of such columns were quite low, but labor costs were extremely high. Most of the developmental work in conventional glass capillary columns occurred in government and university laboratories, where labor accounted for an appreciable portion of a relatively fixed budget, permitting labor costs to be ignored. A number of such ''centers of excellence,'' which devoted a major portion of their efforts to improving the capillary column and the instrumental hardware associated with it, came into being. In terms of theoretical plate numbers, column efficiencies were often quite high, but aggression toward some solutes could be demonstrated. The most critical point, however, was the fragility of those glass capillary columns, which precluded their more general acceptance for routine analytical applications.

As a prelude to our considerations of column activity and column deactivation, it is probably wise to begin with the cautionary note that although not all activity problems are due to the column, it is most commonly faulted. Sample components can be abstracted by the container and the syringe and syringe needle (*vide infra*); active sites, leading to the disappearance of some components and the tailing (i.e., reversible absorption) of others, often occur in the inlet and in portions of the detector (i.e., the flame ionization detector flame jet). Diagnosis and correction of these problems are discussed in later chapters; our attention will be directed to the column *per se*.

Column activity can be evidenced as (1) total subtraction of a solute, (2) partial disappearance of a solute, or (3) skewed or tailing peaks. These may be engendered by catalytic activity or by adsorptive processes, which may be reversible or irreversible. In coping with these problems, our goals are the production of a truly inert surface, fully wettable by the stationary phase, whose inertness will endure even under continuous high-temperature use. The response of the surface to the deactivating treatment will be influenced by the composition of the surface and by its previous history. Although fused silica is much more uniform in terms of composition, it does exhibit batch-to-batch variations, especially in the abundance of free silanols, hydrogen-bonded silanols, and siloxane bridges and their degree of strain. New untreated tubing also contains contaminants and atmospheric residues, some of which were present in the fused silica blank and some of which were generated from other materials during the high-temperature drawing process. New tubing should be given a careful cleaning, followed by inert gas flushing. The degree of surface hydroxylation on virgin fused silica tubing is usually less than that of conventional glass and varies considerably between lots. In accordance with the considerations discussed below, the tubing is usually given an acidic rinse to hydroxylate the surface and dried under conditions that will ensure the removal of free water and still preserve a maximum number of free silanols, avoiding the creation of strained siloxane bridges.

2.3   Sources of Activity

Conventional Siliceous Glasses

From the combined results of a great many investigators, many of which have been detailed elsewhere [4], a better understanding of the activity displayed by glass columns toward some solutes has finally emerged. The siliceous glasses are based on a lattice composed of alternating silicon and oxygen atoms (Fig. 2.1 [1]). The conventional siliceous glasses, however, also contain appreciable quantities of other substances [1,5,6], some of which occur fortuitously (e.g., metal oxides) while others are added during manufacture (e.g., boron). Metal ions (and boron) are characterized by the ability to interact with electron-donor solutes (i.e., many S-, N-, and O-containing compounds). In addition, the tetravalent silicon atoms exposed at the surface of the siliceous matrix satisfy some of their nonlattice bonds with —OH (silanol) groups, which can also serve as Lewis acid sites; all of these lead to activity problems.

The combined contributions of many investigators have led to methods for leaching these nonsiliceous materials (largely metal oxides and boron) from the

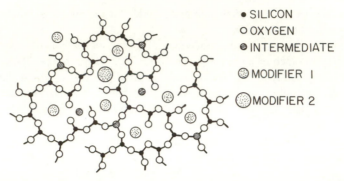

Fig. 2.1. Diagrammatic representation of the structure of conventional siliceous glasses. The structure is based primarily on a silicon–oxygen lattice; inclusion of other atoms disrupts the regularity of the lattice. (Adapted from [5]; reprinted from W. Jennings, ''Comparisons of Fused Silica and Other Glass Columns for Gas Chromatography'' (1981) with permission of Huethig Verlag, Heidelberg, New York.)

surface of the glass capillary (see, e.g., [7–9]). However, only the active sites that occur at the glass surface are available for removal by these leaching processes; potentially active sites buried in the bulk glass remain unaffected. These siliceous glasses, of course, are not solids, but highly viscous liquids. Although diffusion does occur, it is normally very slow because of the high viscosity of the medium. As the glass temperature is increased, the viscosity of the medium drops and diffusivity increases [10]. Flame straightening of column ends accelerates the diffusion of ions buried within the matrix, some of which reach the surface and restore it to an active state. These various defects seriously limit the utility of conventional glasses for open tubular columns.

Quartz and Fused Silica Glasses

The average chromatographer faulted conventional glass capillaries on two points: (1) they were fragile and prone to breakage, and (2) they exhibited excessive activity, i.e., deactivation was a problem. Later studies (some of which are reviewed below) established meticulous procedures by which the surface of the conventional glass capillary could be rendered reasonably inert. However, ideal chromatographic connections usually require straightened column ends, and heating conventional glass to the point that it can be straightened has an adverse effect on the longevity of that deactivation treatment. In 1979, Dandeneau and Zerenner [11] investigated the chromatographic suitability of several glasses, including fused silica. It should be noted that although the words ''quartz'' and ''silica'' are sometimes used almost interchangeably (particularly in column advertisements), the two materials are different; both are higher-purity siliceous glasses. Naturally occurring quartz (principally from Brazil and

Madagascar) is processed commercially to produce a fused quartz. This material is available in ingots, bars, and tubing and contains about 100 ppm metallic oxides, dominated by those of aluminum and iron. Several refined grades of fused quartz are also commercially available; refining generally consists of subjecting the finely ground quartz powder to acid leaching and remelting. In some grades, the metallic oxide content has been reduced to about 10 ppm. The term "fused silica" is usually reserved for a synthetic product; under "clean-room" conditions, high-purity synthetic silicone tetrachloride is introduced as a fine spray into a hydrogen flame, where the $SiCl_4$ decomposes and combines with the water produced by combustion of the hydrogen to yield silicon dioxide, SiO_2, and HCl. The former is collected electrostatically and melted to produce fused silica. Fused silica, which can be regarded as a synthetic quartz, can be produced with a metallic oxide content of less than 1 ppm. Both fused quartz and fused silica are used in column manufacture. Column manufacturers must also pay considerable attention to other variables, such as the level of free hydroxyl (silanol) in the quartz or fused silica, and imperfections such as surface flaws that may occur in the tubing blanks.

Dandeneau and Zerenner [11] recognized that the tensile strength of this material was immense, because its higher purity resulted in a more orderly structure with a higher degree of cross-linking. This increased strength permitted the construction of exceedingly thin-walled columns (0.2 mm i.d. × 0.25 mm o.d.) that were inherently straight, extremely strong, and highly flexible. The more orderly cross-linked structure also necessitated higher working temperatures; the much more expensive blanks of these high-purity materials undergo softening somewhere above 2000°C. The lifetime of a conventional oven under these conditions would be relatively short; a flowing argon atmosphere is usually provided, and, even here, vaporization of metal ions from the heated apparatus and deposition of those ions on the freshly drawn tubing can lead to serious problems. Fused silica and quartz (*vide infra*) also have very narrow ranges of suitable "working temperature"; another complication is that even water vapor and airborne dust can cause flaws in this high-purity tubing that lead to breakage. Because of these various considerations, it is desirable that drawing proceed as rapidly as possible once drawing conditions are idealized. Fused silica is typically drawn at rates that may approach 1–2 m/sec. These restrictions make imperative the use of more sophisticated drawing machinery, operating under clean-room conditions and based on fiber-optic technology.

These considerations lead to the conclusion that, as compared to conventional glass capillary tubing, fused silica is both material-intensive and labor-intensive; it is an expensive product. Many of the "centers of excellence" that contributed so much to the understanding and development of capillary gas chromatography were effectively priced out of the market by the introduction of fused silica capillaries; to some degree, this situation has improved.

2.4   Structural Flaws

All glasses have surface flaws, and fused silica is no exception. Many such defects survive the drawing process to produce flawed areas in the drawn tubing, with adverse effects on its strength (Fig. 2.2). To minimize these flaws, the fused silica blanks used in column production are usually subjected to careful selection and then meticulously fire polished. Even so, some flaws persevere to produce weakened points in the drawn tubing. In addition, particulate matter (e.g., dust) that comes in contact with the tubing during these early phases of its manufacture will result in surface flaws; drawing must take place under conditions of extreme cleanliness.

Michalske and Freiman [12] reported that the siliceous glasses undergo a time-related decrease in mechanical strength which is encouraged by static load and certain environmental factors. This structural weakening is due to a stress corrosion process consisting of a growth (usually slow) of preexisting surface flaws. Their data indicate that the velocity of flaw extension (i.e., crack growth) is related to the stress intensity and is encouraged by liquid water, moist gas, or ammonia. Possible because of the more orderly lattice, fused silica (and quartz) seems especially susceptible.

Conventional glass capillaries were forced through a curved tubular furnace (bending tube) to produce a continuous coil [4]; interfacing the finished column to inlets and detectors usually required straightening the column ends. Fused silica is drawn as a long straight tube, which is then forced to a coiled configuration. The finished column is much easier to interface to inlets and detectors, but the coiled column is under stress. The degree of stress varies directly with the radius of the tubing and inversely with the coiling diameter (Fig. 2.2).

To inhibit these surface corrosion processes, the thin-walled flexible fused silica tubing is drawn through a reservoir of polyimide solution as it emerges from the drawing oven. The coating thus deposited on the outer tubing wall is imperfect and has pinholes. The freshly coated tubing is usually drawn through a tubular drying oven, recoated, and redried before the drawing rollers can come in contact with it. In some cases, it is then subjected to additional coating treatments [13]. The polyimide coating probably serves two functions: (1) it floods and fills existing flaws, which tends to discourage their continued growth, and (2) it seals the outer surface of the capillary with a waterproof barrier, limiting corrosive attack by, e.g., water vapor. For reasons that are not completely understood, the inner surface of the tubing is much less susceptible to these corrosion problems.

In attempts to detect flawed tubing so that it can be discarded as early as possible in the manufacturing process, most manufacturers first subject the tubing to dynamic and static (fatigue) tests. The flaws that survive these testing procedures and the column manufacturing processes are usually minor and cause no problems. On rare occasions, user-sustained breakage can be traced to flaws

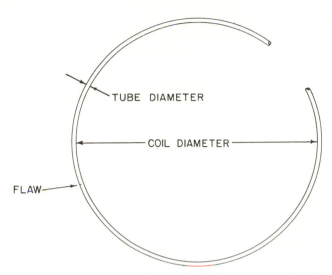

Fig. 2.2. Diagrammatic representation of factors affecting the growth of surface flaws in siliceous glasses; see text for details.

that should have been revealed by the above procedures, but breakage in the hands of the user more commonly results from abrasion or scratching of the outer wall. This violates the integrity of the polyimide film and permits corrosion at that point.

While the polyimide has excellent mechanical resistance and good temperature tolerance, it also has a tendency to render the column opaque. When first applied, the better-quality material has a light golden color and is semitransparent. At temperatures in excess of 300°C, it becomes darker; the darkening is encouraged by oxygen and is apparently influenced by batch-to-batch differences in the polyimide. In most cases, we would prefer a transparent column; not only does it make possible the discernment of localized defects in the column, but also it can be advantageous in certain injection modes (see Chapter 3). Some developments in pyrocarbon coatings are rumored, and there are also reports of thermally stable transparent coatings [14]. Aluminum and nickel coatings have been applied to fused silica tubing by vapor phase deposition, but the cost of the completed tubing is significantly higher.

It is also possible to apply an outer metal sheath over the column; this can be accomplished by drawing the uncoated fused silica tubing through a bath of molten aluminum [15]. There are some reports that breakage of the fused silica inside this sheath, which would be difficult to detect, is not uncommon; the molten aluminum would be less effective at flooding and sealing existing flaws in the surface. The major utility of such columns would appear to be for extremely high temperature use (e.g., 450–500°C), where polyimide coatings would be entirely unsatisfactory. The upper feasible temperatures for gas chromatography, however, are imposed not only by the outer coating on the column, but also by the temperature-dependent degradation of stationary phases (see below) and, most important, by the thermal stability of the solutes themselves. Few compounds other than hydrocarbons can tolerate these extreme temperatures; with other solutes, the analyst must be concerned that peaks may be due to pyrolysis products of the parent solute.

The temperature-related stresses to which the column is subjected become much more severe at these extreme temperatures, and column lifetimes may be considerably shortened. Stationary phases are in general subject to oxidative degradations that are both oxygen- and temperature-dependent: oxygen levels so minute that they may be tolerable at 300°C lead to abbreviated column lifetimes at extremely high temperatures (see Chapters 4 and 10). At our present state of development, these very low volatility solutes are probably better analyzed by liquid or supercritical fluid chromatography.

2.5 Flexible Columns of Conventional Glasses

Consideration should also be given to another aspect which has been a source of confusion for some chromatographers. Certainly the blanks for drawing either quartz or fused silica capillaries are much more expensive than blanks of conventional glass, and the former require much more expensive drawing machinery and much more exacting processes; because of these constraints, quartz and fused silica capillary tubing are much more expensive than conventional glass capillary tubing. Conventional glasses can also be drawn to very thin-walled capillaries, and there are reports that by coating such tubing with polyimide, one can produce less expensive columns that have the other advantages of quartz and fused silica (e.g., [16]); most authorities regard those claims as exaggerations. The use of various conventional glasses to construct thin-walled flexible columns is one of the many alternative routes that column manufacturers have exhaustively explored and then abandoned. The fact that such tubing is weaker, less flexible, and more subject to static fatigue might be tolerable, but there is a more serious defect: columns produced from that tubing cannot be well deactivated, because they cannot be leached without making the tubing extremely fragile [17]. Consequently, one of the archaic and thermally labile treatments for masking the metal oxides must be used for their deactivation. The net result is a less flexible, more fragile column that is less inert (indeed, it can be extremely active) and is restricted to lower operating temperatures [18]. Conventional glass capillaries have also been coated internally with an inner envelope of polyimide prior to coating with the stationary phase [19]; again, the polyimide-coated surface requires masking prior to application of the stationary phase, and the utility of such columns appears to be questionable.

2.6 Silanol Deactivation

The results originally reported by Dandeneau and Zerrener [11], supported by many subsequent reports, indicated that fused silica columns are in most cases far more inert than columns of conventional glasses. Column activity is due to both metal oxides and surface silanols (plus strained siloxane bridges) [1]. In conventional glass columns, both contribute to chromatographic defects, and both require attention. As discussed earlier, metal contaminants in the surface of conventional glass columns are generally removed by leaching, rinsing, and careful drying, before attention is directed to the surface silanols. Fused silica has essentially no metal contamination, and the problem of deactivation is reduced to one of dealing with surface silanols; there are some complicating factors. Because of the difference in the temperatures required for their drawing, conventional glasses possess a number of strained siloxane bridges [1] which readily rehydrate to silanols, while fused silica is less hydroxylated. Fused silica blanks can be obtained with high or low hydroxyl content, and drawn tubing can be readily rehydroxylated by washing with aqueous acids (e.g., [20]).

Earlier methods for silanol deactivation have been reviewed elsewhere (e.g., [4,21]); our treatment here will be abbreviated. Because it had been used successfully to deactivate silanol groups on siliceous packings, silylation, primarily with the chlorosilanes, was one of the earliest deactivation procedures investigated for glass capillaries (e.g., [22]). However, used either solely for the deactivation of glass capillaries or jointly for their deactivation and surface modification (by bonding other functional groups to the surface), these procedures usually produced columns that were faulted for excessive activity and/or low efficiency. The first encouraging silylation results employed hexamethyldisilazane (HMDS) at 300°C for up to 20 hr [23]. Grob *et al.* [7] reasoned that many of the inconsistencies subsequently reported by a variety of authors could be corrected by first resorting to a leaching process, which served two purposes: (1) dissolution and removal of metal oxides from the glass surface (the aggressiveness of these active sites would not be lessened by silylation) and (2) modification of the silica structure so that it presented an ''optimum number'' of silanol groups (i.e., the surface should be fully hydroxylated). One critically important step, then, is that following leaching: water, which would reduce the efficacy of silylating reagents and also react with disilazanes to produce high concentrations of ammonia, must be removed; hydrogen-bonded silanols should remain as free silanols and not be driven to strained siloxane bridges. Wright *et al.* [24] also explored the use of chlorosilanes for high-temperature deactivation. Schomburg *et al.* [9] demonstrated that *in situ*-produced thermal decomposition products of polysiloxane phases effectively neutralized active sites; Stark *et al.* [25] reasoned that these decomposition products must be siloxane fragments and introduced the concept of using cyclic trimers and tetramers of —$Si(CH_2)_2$—O— in high-temperature vapor phase deactivation. Superior results were achieved with octamethyltetracyclosiloxane; hexamethyltricyclosiloxane was less effective. The idea was further developed by Blomberg et al. [26,27]. Lee [28] reported that at 420°C the former reagent (tetramer) blanketed the fused silica surface with a thin film of glassy polysiloxane resin, in which localized pools of resin were evident. They suggested that this polymeric resinous coating was essential to effective deactivation and that simple silylation of surface silanols, which would be expected to occur at more moderate temperatures, was insufficient. They also reported that deactivation treatments utilizing the chlorosilanes were generally inferior to those employing either the cyclic siloxanes or polysiloxanes.

2.7 Column Coating

Stationary phases are discussed from the standpoint of "polarity," functionality, diffusivity, and other factors in Chapters 4 and 5; at this point our concern is confined to generalities of the coating operation. The goal, of course, is to deposit on the inner periphery of the cleaned, deactivated tubing a thin, uniform film of stationary phase. The role of many interrelated factors and the several methods used for coating stationary phases have been reviewed elsewhere [4]; the static technique (with various minor modifications) is most widely used today.

The stationary phase may endure on the column wall because (1) cohesive or wetting forces exist, (2) the stationary phase in contact with the surface may be bonded to the surface to produce a stable monolayer, wetted by the remainder of the stationary phase, (3) it may be cross-linked to produce a liner or envelope of stationary phase polymer within the column, or (4) it may be both cross-linked and surface-bonded. Cross-linking can produce a stationary phase that is nonextractable under normal use conditions; this can offer distinct advantages for use with either splitless or on-column injection, where the column is flooded with liquid. It is also advantageous where the column has become contaminated with soluble but nonvolatile or high-boiling residues (as in on-column injection), as it enables these to be washed from the column with a suitable solvent (see later chapters). Surface bonding is advantageous for "superthick" (i.e., 1.5–5 μm) film columns, in which the envelope of stationary phase can otherwise collapse and occlude the column [29]. Cross-linked surface-bonded phases are also desirable for columns used with supercritical fluid and liquid chromatography.

Surface bonding has been employed to attach stationary phases to packings, but this approach has not proved useful in capillary chromatography. The earliest results probably date to Bossart [30] and the polybutadiene layer bonded by Grob [31]. Because they utilized raw, untreated glass surfaces, the latter efforts produced columns which, while interesting, had low thermal stability. Later developments have made it clear that surface preparation is an integral part of the bonding process. Further developments can be traced to attempts by Blomberg and Wannman [32] and by Grob [33] and to the initial discovery by Jenkins [34] (see Chapter 4). Additional efforts that contributed to these developments, consideration of which is beyond the scope of this book, appear in references [35–39]. Grob's recent book [40] will prove especially valuable to those who wish to manufacture their own columns.

Details of stationary phase properties and other facets of the bonding and cross-linking reactions are discussed later in Chapter 4; in general, the initiators used to trigger free-radical reactions resulting in C–C cross-linking can be divided into peroxides (e.g., [41–47]), azo compounds (e.g., [48,49]), gamma radiation from ^{60}Co [50–52], accelerated electrons from a Van de Graaff generator [53], and ozone [54]. Because the products of their decomposition are volatile, dicumyl peroxide (e.g., [41,43]) and azo compounds (e.g., [48]) are widely used. Ionizing radiations have been faulted for causing a loss of flexibility in fused silica columns [55].

A final note, based on more than a decade of experience in the commercial manufacture of many thousands of columns, seems in order. Progress in the preparation of high-efficiency columns of increased inertness has been excellent. Most of our knowledge is based on the accumulation of bits and pieces of information from a variety of different investigations. Evaluation of this evidence requires the reconciliation of apparent contradictions and discounting of some claims that seem exaggerated. Most of these reports have been concerned with results obtained on one to several columns, and those samples are far too small for the law of averages to become evident. Results obtained with larger lots sometimes indicate that many of the results reported would appear to reflect an exception rather than the rule.

References

1. W. Jennings, "Comparisons of Fused Silica and Other Glass Columns for Gas Chromatography." Huethig, Heidelberg, 1981.
2. W. Jennings, *J. High Res. Chromatogr.* **3**, 601 (1980).
3. D. H. Desty, J. N. Haresnip, and B. H. F. Whyman, *Anal. Chem.* **32**, 302 (1960).
4. W. Jennings, "Gas Chromatography with Glass Capillary Columns," 2nd ed. Academic Press, New York, 1980.
5. J. R. Hutchins, III and R. V. Harrington, *Encycl. Chem. Tech., 2nd ed.* **10**, 533 (1966).
6. G. W. Morey, "The Properties of Glass," 2nd ed. Van Nostrand-Reinhold, Princeton, New Jersey, 1154.
7. K. Grob, G. Grob, and K. Grob, Jr., *J. High Res. Chromatogr.* **2**, 31 (1979).
8. G. Schomburg, H. Husmann, and H. Borowitzky, *Chromatographia* **12**, 651 (1979).
9. G. Schomburg, H. Hussmann, and H. Behlau, *Chromatographia* **13**, 321 (1980).
10. R. H. Doremus, "Glass Science." Wiley, New York, 1973.
11. R. Dandeneau and E. Zerenner, *J. High Res. Chromatogr.* **2**, 35 (1979).
12. T. A. Michalske and S. W. Freiman, *Nature (London)* **295**, 511 (1982).

13. S. R. Lipsky, W. J. McMurray, M. Hernandez, J. E. Purcell, and K. A. Billeb, *J. Chromatogr. Sci.* **18**, 1 (1980).
14. C. DeLuca, personal communication (1983).
15. R. Dandeneu, personal communication (1982).
16. R. P. W. Scott, *TrAC, Trends Anal. Chem. (Pers. Ed.)* **2**, 1 (1983).
17. S. R. Lipsky, *J. High Res. Chromatogr.* **6**, 452 (1983).
18. W. Jennings, *Am. Lab.* **16** (1), 14 (1984).
19. J. Balla and M. Balint, *J. Chromatogr.* **299**, 139 (1984).
20. M. W. Ogden and H. M. McNair, *J. High Res. Chromatogr.* **8**, 326 (1985).
21. M. L. Lee, F. J. Yang, and K. D. Bartle, "Open Tubular Column Gas Chromatography: Theory and Practice." Wiley (Interscience), New York, 1984.
22. M. Novotny and K. Tesarik, *Chromatographia* **1**, 332 (1968).
23. T. Welsch, W. Engwald, and C. Klaucke, *Chromatographia* **10**, 22 (1977).
24. B. W. Wright, M. L. Lee, S. W. Graham, L. V. Phillips, and D. M. Hercules, *J. Chromatogr.* **199**, 355 (1980).
25. T. J. Stark, R. D. Dandeneau, and L. Mering, *Pittsburgh Conf. Anal. Chem. Appl. Spectrosc., 1980,* Abstract 002 (1980).
26. L. Blomberg, K. Markides, and T. Wannman, *J. High Res. Chromatogr.* **4**, 527 (1980).
27. L. Blomberg, K. Markides, and T. Wannman, *Proc. Int. Symp. Capillary Chromatogr., 4th, 1981,* p. 73 (1981).
28. M. L. Lee, R. C. Kong, C. L. Wooley, and J. S. Bradshaw, *J. Chromatogr. Sci.* **22**, 136 (1984).
29. K. Grob and G. Grob, *J. High Res. Chromatogr.* **6**, 133 (1983).
30. C. J. Bossart, *ISA Trans.* **7**, 283 (1968).
31. K. Grob, *Helv. Chim. Acta* **51**, 729 (1968).
32. L. Blomberg and T. Wannman, *J. Chromatogr.* **168**, 81 (1979).
33. K. Grob, *Chromatographia* **10**, 625 (1977).
34. R. G. Jenkins and R. H. Wohleb, *Pap., Expochem 80,* (1980).
35. K. Grob and G. Grob, *J. Chromatogr.* **213**, 211 (1981).
36. K. Grob and G. Grob, *J. High Res. Chromatogr.* **5**, 13 (1982).
37. T. J. Stark and P. A. Larson, *J. Chromatogr. Sci.* **20**, 341 (1982).
38. P. A. Peaden, B. M. Wright, and M. L. Lee, *Chromatographia* **15**, 335 (1982).
39. L. Blomberg, J. Buijten, K. Markides, and T. Wannman, *J. Chromatogr.* **239**, 51 (1982).
40. K. Grob, "Making and Manipulating Glass Capillary Columns." Huethig, Heidelberg, 1986.
41. J. Buijten, L. Blomberg, K. Markides, and T. Wannman, *Chromatographia* **16**, 183 (1983).
42. P. Sandra, M. Van Roelenbosch, I. Temmerman, and M. Verzele, *Chromatographia* **16**, 63 (1983).
43. K. Markides, L. Blomberg, J. Buijten, and T. Wannman, *J. Chromatogr.* **254**, 53 (1983).
44. K. Grob, G. Grob, and K. Grob, Jr., *J. Chromatogr.* **211**, 243 (1981).
45. P. Sandra, G. Redant, E. Schacht, and M. Verzele, *J. High Res. Chromatogr.* **4**, 411 (1981).
46. L. Blomberg, J. Buijten, K. Markides, and T. Wannman, *J. High Res. Chromatogr.* **4**, 578 (1981).
47. J. Buijten, L. Blomberg, K. Markides, and T. Wannman, *J. Chromatogr.* **268**, 387 (1983).
48. B. E. Richter, J. C. Kuei, N. J. Park, S. J. Crowley, J. S. Bradshaw, and M. L. Lee, *J. High Res. Chromatogr.* **6**, 371 (1983).
49. B. E. Richter, J. C. Kuei, J. S. Bradshaw, and M. L. Lee, *J. Chromatogr.* **279**, 21 (1983).
50. G. Schomburg, H. Husmann, S. Ruthe, and M. Herraiz, *Chromatographia* **15**, 599 (1981).
51. W. Bertsch, V. Pretorius, M. Pearce, J. C. Thompson, and N. Schnautz, *J. High Res. Chromatogr.* **5**, 432 (1982).
52. J. A. Hubball, P. R. Dimauro, S. R. Smith, and E. F. Barry, *J. Chromatogr.* **302**, 341 (1984).

53. K. Markides, L. Blomberg, J. Buijten, and T. Wannman, *J. Chromatogr.* **267,** 29 (1983).
54. J. Buijten, L. Blomberg, S. Hoffmann, K. Markides, and T. Wannman, *J. Chromatogr.* **289,** 143 (1984).
55. J. A. Hubball, P. Dimauro, E. F. Barry, and G. E. Chabot, *J. High Res. Chromatogr.* **6,** 241 (1982).

CHAPTER 3

SAMPLE INJECTION

3.1 General Considerations

Component separation is affected both by the ratio of solute retentions (α) and by the width of the solute peaks (Chapter 1). Peak widths are influenced by, among other things, the efficiency of the injection process: the shorter the length of column occupied by the injected sample, the shorter the band as it begins the chromatographic process, the shorter the band as it completes the chromatographic process and is delivered to the detector, the sharper the peak generated by the detector, and the better the separation. One of the most critical functions of the injection process is to introduce to the chromatographic column a starting band of sample that is as short as possible.

Solute bands lengthen, inevitably and inexorably, during the chromatographic process. In the ideal case, the increase in length would result only from longitudinal diffusion in the mobile phase; in actuality, it is usually influenced by other factors, such as active sites that delay or prolong the passage of some but not all of the molecules of a given solute, or by irregularities in the stationary phase coating. These phenomena would increase the degree of randomness in the behavior of identical solute molecules and lead to lengthened and more dilute bands. Longer bands produce broader peaks, which have adverse effects on component separation.

To assist the reader to more easily correlate our discussion with presentations in the literature, it should be pointed out that the above phenomenon is often referred to as "band broadening," probably because the end product is broadened peaks. The transverse dimension of the band (i.e., bandwidth) is fixed by the column; it is the longitudinal dimension that varies, and it is really the length of the band that is under consideration.

It is also important that the qualitative and quantitative composition of the injected band accurately reflect that of the original sample and that it contain detectable amounts of each substance to be detected. Factors influencing the degree to which this is accomplished include

1. discriminatory effects during transfer of the sample from the syringe to the inlet (these are usually, but not always, of greater concern in the split and splitless modes);

2. discriminatory effects during transfer of the sample from the inlet to the column (for one thing, inlets often abound in active sites);

3. the length of column initially occupied by the injected sample (all modes);

4. the rapidity of sample vaporization in the inlet (vaporizing modes, i.e., split, splitless);

5. the speed of sample transport from the inlet to the column (split, splitless);

6. the completeness of sample transport from the inlet (all modes);

7. the homogeneity (or heterogeneity) of (a) the temperature and (b) the phase ratio (both real and "apparent," i.e., condensed liquid) throughout the entire length of the sample band, i.e., "distribution constant" and "phase ratio" effects (*vide infra*).

Many facets of the injection process have previously been considered in some detail [1], and our treatment here will be largely supplementary. Different procedures are usually employed for injections into large-diameter (i.e., greater than 0.35 mm i.d.) columns, as contrasted with small-diameter (i.e., less than 0.35 mm i.d.) open tubular columns. Some basic considerations apply to both.

3.2 Extrachromatographic Phenomena Influencing Band Length

Syringe Technique

Discriminatory effects occasioned by the syringe have been considered elsewhere; briefly, filled-needle injections or plunger-in-needle syringes are almost always undesirable in the split or splitless capillary modes. Split mode requires a fast injection, splitless usually gives better results when the "hot needle" technique [1–3] is employed. Under some conditions, other facets of the syringe technique are critically important to the chromatographic results obtained; this is particularly true with on-column injections, where the syringe needle deposits the sample inside the chromatographic column. Rapid depression of the plunger on a standard 10-μl syringe expels the sample with considerable velocity; in free air, the unrestrained ballistic path can distribute a spray pattern of material over a distance exceeding 1.5 m. With the needle housed inside a column, that distance would be much less, because collision of the spray with the column wall would impose limits on the length of column subjected to the spray pattern; the linear movement of the mobile phase would complicate these considerations. Chromatographic defects such as peak splitting and peak broadening can be generated by too rapid injection of sample volumes greater than ~1.0 µl. Too slow an injection, on the other hand, may result in wetting of the outside of the needle, and capillarity can then draw sample into the concentric space between the needle and the column; that band may be smeared still further as the needle is withdrawn from the column.

Liquid-Induced Spreading

Even in the case of an "ideal" on-column injection, where a short, discrete band is deposited on the coated column, problems may develop. The injected band of mixed solutes and solvent is not stationary, but advances through the column under the impetus of the mobile phase, albeit at a lower velocity. As it advances, solutes diffuse from the moving solvent–solute band to the stationary phase, smearing sample over an extended length of column; this phenomenon has been termed "band broadening in space." If the injected plug should "bridge" the column, the velocity of the sample plug will be that of the mobile phase, and the band will be smeared over an even longer section of column. The term "band broadening in space" was used to distinguish this spatial phenomenon from "band broadening in time" [4,5]. The latter can be envisioned as a temporal (time-dependent) band-lengthening process occasioned by delivery of the sample to the column slowly and over a period of time; before and during transport of the latter portions of the sample, portions deposited earlier have already begun their chromatographic development. This can be caused by (among other things) slow volatilization of the sample in the inlet and/or excessive inlet volume relative to the gas flow (split or splitless). Other aspects of these band-lengthening processes are considered later in this chapter.

3.3 Chromatographic Factors Influencing Band Length

Following the deposition of the sample on the column, the length of column occupied by the band can be further affected; this can have positive or negative effects on the chromatographic results finally obtained. Equation (1.14) can be rewritten as

$$k = K_D/\beta = c_S V_S/c_M V_M \qquad (3.1)$$

The velocity of a chromatographing band varies directly with the velocity of the mobile phase and indirectly with the proportion of the analysis time that the solute spends in stationary phase, i.e.,

$$v = u/(k + 1) \qquad (3.2)$$

Hence, the velocity of a solute band can be expressed as

$$v = [u(\beta/K_D)] + 1 = [u(c_M V_M / c_S V_S)] + 1 \qquad (3.3)$$

If the velocity is identical at every point throughout the band, the band moves as a unit whose length is affected only by longitudinal diffusion and by heterogenities in the flow path.

Distribution Constant Effects

If a solute band chromatographs under conditions where the distribution constant K_D is not uniform throughout the band, the velocity is not uniform throughout the band, and the length of the band will be affected; the term "distribution constant effects" has been proposed for these processes [6–8]. The concept is exploited in "cryogenic focusing" and in "thermal focusing" [9,10]. The distribution constant is a function of temperature; ..₃ the temperature decreases, K_D increases, because of the increase in c_S and decrease in c_M. Conversely, an increase in temperature obviously causes a decrease in K_D. In both of the above focusing steps the band progresses from a warmer section of column to a cooler section of column; if a flowing coolant is employed, the direction of flow must be counter to that of the carrier gas. Because $K_{D\ front} > K_{D\ rear}$, $v_{front} < v_{rear}$ and the band is shortened or focused. In "thermal focusing" the band shortening phenomena are further exploited in the subsequent heating step, as heat is applied from the rear of the band toward the front (i.e., in the same direction as the carrier flow). Again, $K_{D\ rear} < K_{D\ front}$, $v_{front} < v_{rear}$, preserving or even enhancing the focus. "Secondary cooling," as it was originally employed in on-column injections [11], normally resulted in lower separation efficiencies, which were in part attributable to a distribution constant defocus that resulted from the positive temperature gradient [12]. Note that temperature programming does not normally exercise these effects: where the temperature is changed over the entire length of the band simultaneously, the c_S/c_M ratios are decreased, but those decreases occur uniformly and simultaneously over the entire band; the entire band is accelerated as a unit. Distribution constant effects have also been exploited in coupled columns and in multidimensional gas chromatography. By using the column of greater polarity (higher distribution constant) last, bands are shortened as they enter that column [13]. The retention gap (see below) also utilizes a distribution constant focus.

Phase Ratio Effects

If a chromatographing band encounters conditions where the phase ratio β experienced by one section of the band differs from that experienced by other sections of the band, localized band velocities are not uniform and the length of the band is affected; the term "phase ratio effects" has been proposed for this phenomenon [7–9]. One classical example is the "solvent effect" (e.g., [14–

16]), in which a more volatile solvent is condensed on the column just prior to the time when the less volatile solutes move into that area. The condensed solvent changes the apparent phase ratio of the column, and the front of the solute bands encounters a lower phase ratio than does the rear of the solute bands. Hence $v_{front} < v_{rear}$, and the band is shortened or focused. The phenomenon can also be observed when a major solute is present in amounts so great that it effectively changes the "apparent phase ratio" of the column for solutes in the immediate vicinity of that major component [17]. This concept, too, has been used in multidimensional gas chromatography, by following a standard film column with an ultrathick film column to achieve band shortening without cold trapping [13]. These phenomena are exploited in a variety of ways, sometimes fortuitously and sometimes deliberately; the retention gap (see below) also employs a phase ratio focus.

3.4   Injection into Large-Diameter Open Tubular Columns

In conventional (low-flow) capillary columns, the inlet may require modification so that a high gas flow can rapidly flush the vaporized sample from the

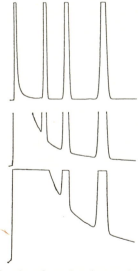

Fig. 3.1. Solvent tailing as a function of sample volume, using a "normal" liner and vaporizing injections on a Megabore column operated in high flow mode. Hydrocarbons (C_{11}, C_{12}, C_{13}) in hexane; inlet temperature 250°C; helium carrier at 30 ml/min. Top, 0.5 μl; center, 1.0 μl; bottom, 2.0 μl. The rapidly vaporized sample "overflows" the liner (flashback) and is forced out of the main flow stream. As the pressure pulse decays, solvent vapor diffuses back into the flow stream, producing the solvent tail. (Adapted from the *Journal of Chromatographic Science,* volume 24, 1986, p. 34 [7], with permission of the copyright holder, Preston Publications, Inc.)

injection chamber in spite of the restricted flow through the column. A major advantage of the larger-bore (e.g., 0.53 μm i.d.) open tubular columns is that the standard packed column injector can be used without any special provisions for stream splitting. The carrier gas flow rate through these columns can be the same as that used in packed columns; as a result, the standard injector is rapidly cleansed. However, other injection-related problems can be encountered.

Figures 3.1–3.3 illustrate that direct vaporizing injections into these large-diameter open tubular columns can create problems when used with the normal straight-bore inlet liner. Larger injections, or slower (more nearly optimal) carrier gas flows, often result in severe tailing of the solvent and more volatile solutes and discrimination against some higher-boiling solutes [7]. All of these problems are attributable to backflash; the severity of backflash would be expected to increase with increasing injection size and with decreasing flow

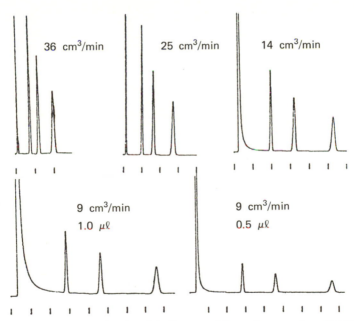

Fig. 3.2. Solvent tailing as a function of the flow rate of the carrier gas, using a "normal" liner and vaporizing injections on a Megabore column. Conditions as detailed in Fig. 3.1 except as noted below. Top chromatograms: 1.0 μl injections at: top, 36 ml/min; center, 25 ml/min; right, 14 ml/min. Bottom chromatographs, helium carrier at 9 ml/min; left, 1.0 μl injection; right, 0.5 μl injection. Where the size of the injected sample is enough to just fill (or slightly overfill) the liner, higher flows are more effective at rapidly transporting the sample to the column (see first four chromatograms). At low flows, the sample size must be decreased if problems are to be avoided with the "normal" liner (bottom chromatograms). (Adapted from the *Journal of Chromatographic Science,* volume 24, 1986, p. 34 [7], with permission of the copyright holder, Preston Publications, Inc.)

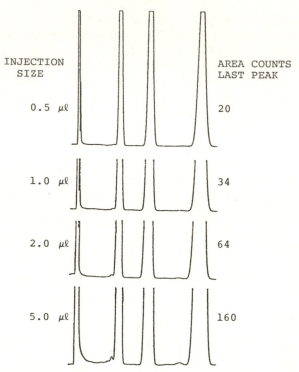

Fig. 3.3. Effect of injection size on quantitative delivery of solutes, using a ''normal'' liner and vaporizing injections on a Megabore column. Helium carrier at 30 ml/min; other conditions as detailed in Fig. 3.1, except for injection size. Flashback, the severity of which correlates with the size of the injection, forces sample (i.e., solvent plus solutes) out of the liner area. Components then condense on cooler surfaces (e.g., the septum face); losses are greater for higher-boiling components. (Adapted from the *Journal of Chromatographic Science,* volume 24, 1986, p. 34 [7], with permission of the copyright holder, Preston Publications, Inc.)

through the inlet and to be exacerbated as the differential between the temperature of the injector and the boiling point of the solvent (and solutes) became greater. Where higher-boiling solutes are forced against the face of the cooler septum, some of those solutes condense and are lost to the analysis. These considerations led to the design of a new injector liner for the large-diameter open tubular column [18].

Figure 3.4 shows two types of liners recommended for use with large-diameter open tubular columns; that shown on the right is for flash-vaporized injections and is preferred in the vast majority of cases. The annular space between the inner wall of the upper constriction and the syringe needle is very slight, which forces the entering carrier gas to a very high velocity, greatly lessening the possibility of flashback. The end of the column should be cut square and seated

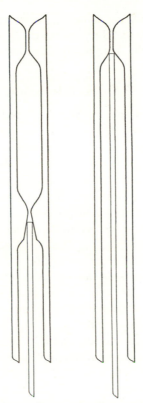

Fig. 3.4. Improved inlet liners for use with Megabore columns. Left, liner for direct flash injection, designed to seal to the square-cut column end. The carrier gas flowing through the restriction formed by the inserted syringe needle is forced to an extreme velocity, effectively preventing flashback. Good results can be obtained with liquid injections as large as 8–10 μl. Right, liner for hot on-column injections. This vaporizing mode should not be confused with cold on-column injection; injection size and the speed of injection must be correlated and are limited by the volume of column receiving the sample. (Adapted from the *Journal of Chromatographic Science*, volume 24, 1986, p. 34 [7], with permission of the copyright holder, Preston Publications, Inc.)

snugly in the ground taper which forms the lower restriction. Vaporization of the sample still results in a pressure pulse, but as long as the syringe needle is in position, the column offers the path of least resistance; instead of flashback, the sample is forced forward into the column. The syringe needle should be left in position until the pressure pulse is dissipated by flow into the column; this is approximately 1 sec per microliter injected. Solvent tailing provides a useful indication of too early syringe needle withdrawal [18].

The other liner illustrated in Fig. 3.4 is used for on-column injections into the large-diameter open tubular columns, as the syringe needle is guided directly inside the column. Some practitioners pack the longer portion with glass wool,

Chromosorb, or similar materials and use it in the inverted position to protect the column during the injection of "dirty" samples. Nonvolatile materials then tend to remain in the inlet, which is periodically removed for cleaning or replacement.

3.5   Split Injection

This injection mode is usually restricted to capillary columns and has been discussed in detail elsewhere [1]. Briefly, the sample is introduced into a heated injector, where it undergoes rapid vaporization and thorough mixing of the volatilized components. Most of the gas flowing through the injector, together with the major portion of the volatilized sample, is vented to the atmosphere, while a minor fraction of each is directed to the column. Because the inlet end of the column is coupled to or lies within the heated injector and the remainder of the column is housed in the cooler oven, the sample chromatographs through a negative temperature gradient: $K_{D\ front} > K_{D\ rear}$, $v_{front} < v_{rear}$, and the band is shortened or focused. Note that at the detector end of the column, the reverse situation prevails; solute bands ascend a positive temperature ramp. Because of the velocity gradient in the column (mobile phase velocity is highest, and the band is moving most rapidly at the outlet end of the column), the effect is usually very slight, but it can be demonstrated.

Split is a vaporizing injection mode, and it is desirable that the volume of the injection chamber be sufficient to accommodate the vaporized sample without engendering either a severe "pressure pulse" or "backflash" of the solvent (and solutes) into remote areas of the injector, against the face of the cooler septum, or back into the carrier gas line. These phenomena not only can affect split ratios and linearity but also can lead to "ghost peaks" and tailing. At the same time, the volume of the injection chamber should not be so large that an inordinate length of time is required to flush the sample from the inlet. Capillary columns of "standard" dimensions are restricted to relatively low volumes of gas flow; yet the injection chamber must be flushed clean almost instantaneously to avoid the introduction of an elongated band of sample.

The apparent contradiction in these requirements is resolved by providing an inlet whose volume is large enough to accommodate the vaporized sample and which is swept by a relatively large gas volume. Only a small portion of that sample-laden gas is conducted to the column; the major portion is vented to atmosphere (i.e., "split") through some type of restriction. The primary function of a splitter is not merely to limit the sample size; far more important is the fact that the splitter permits a high flow rate through the injection chamber, so that the carrier gas following the sample into the column is pure carrier and not exponentially diluted sample.

Another aspect that should be considered here is that split injections require a very high rate of sample vaporization. If vaporization occurs only slowly, the band will be introduced into the column over an extended period of time, even if the vaporized sample is immediately transported to the column. The temperature of the injector should be sufficiently high to ensure "instant" vaporization of the solutes [especially true in isothermal runs and of (at least) the more volatile solutes in programmed runs; see below], but the inlet temperature can be too high. Not only does this subject the sample to an excessive and unnecessary thermal shock, but also the explosive vaporization of the solvent accompanying elevated inlet temperatures leads to solvent backflash and can generate excessive solvent tailing (splitters with a sufficient buffer volume are less prone to back-flash problems). In programmed-temperature runs, higher-boiling solute peaks benefit more from a distribution constant focus and are broadened less by slower solute vaporization than are the lower-boiling solutes. As the solutes move from the heated inlet to the cooler column, the ratio K_{front}/K_{rear} is higher for higher-boiling solutes. Vaporization within the inlet should not be so slow, or oven program rates so high, that the band begins the chromatographic process before all of that solute has been delivered to the column.

Because of this "instantaneous" vaporization requirement, split injectors are operated (for a given sample) at a higher temperature than is required by the other injection modes and may be unsatisfactory for thermally labile materials. Rapid vaporization is encouraged by a heat transfer surface of high thermal mass (e.g., glass bead packing), but such surfaces abound in active sites and may lead to rearrangement and degradation of some sample components. The removable glass (or quartz) liner should normally be subjected to deactivation. Untreated glass liners, glass beads, and glass wool expose the sample to both metal oxides and silanols; substitution of quartz liners (and quartz wool) lessens the problem associated with metal ions, but silanols are still present (see Chapter 2). Methods of deactivating inlets are usually based on leaching and high-temperature silylation treatments derived from earlier observations on conventional glass capillaries; a typical procedure is detailed in Chapter 7.

The fidelity with which that portion of the sample split to the column reflects the actual sample composition is also of concern. A number of competent scientists have studied the effects of recognized variables on splitter performance and found that many interrelated factors play roles. In general, the splitter should achieve extremely rapid vaporization of the sample, which should then be thoroughly mixed and be subjected to expansion just prior to splitting. Although many of these interrelationships are still only partly understood, most would agree that the temperature and the degree of mixing are two factors that seem to exercise major effects on the linearity of a well-designed splitter. Even skilled chromatographers sometimes tend to use conditions that force discrimination with split injectors. Split injection subjects the sample to the most severe thermal shock of any of the common injection modes, and higher split ratios (which achieve better mixing) discard most of the sample. Those recognizing these points often have an understandable tendency to use both lower temperatures and lower split ratios. Inlet temperatures that are unnecessarily high can result in the degradation of thermally labile components, and the rapid vaporization of the sample can cause backflash, evidenced as solvent tailing; this can also affect quantitation (discussed later in this chapter and again in Chapter 10). Lower inlet temperatures can improve the chromatographic results, but if the temperature is not sufficiently high, higher-boiling solutes may generate broadened peaks. The split ratio governs the velocity of the injected sample through the mixing zone; too low a split ratio can lead to inadequate mixing and discrimination. To maintain the necessary velocity, the volume of the inlet liner should be reduced for those cases where very low split ratios are required.

3.6   Splitless Injection

The splitless mode of injection was developed largely by the Grobs (e.g., [9,10,19–22]) as an alternative means of injecting samples onto small-diameter open tubular columns. Again, major portions of this subject have previously been covered in detail [1]. Splitless injection is best suited to the analysis of trace levels of higher-boiling solutes that can be injected in a low-boiling solvent, under conditions where the solvent condenses on the column just before the solutes are transported into that region. Solutes move from an area where the "apparent" phase ratio (i.e., v_M/v_S) is high to an area where it is low, i.e., the phase ratio ramp is negative. Referring to Eq. (3.3), the condensed solvent subjects the front of the solute band to a higher v_S/v_M ratio than is experienced by the rear of the band, $v_{front} < v_{rear}$, and the band is shortened or focused; the "solvent effect" is a manifestation of phase ratio focusing.

The solutes are generally presented in dilute solution in this mode of injection; the solvent must be low boiling with respect to the solutes and high boiling with respect to the column at the time of injection. The vaporization chamber has provision for the mobile phase to be introduced under conditions where either (1) the entire flow is through the inlet to the column or (2) the flow stream is divided, part of it entering the column as carrier gas and part backflushing the chamber through a purge valve to remove the last remnants of the injection. The injection is made with the purge valve closed and all flow passing first through the inlet and thence to the column.

Inasmuch as this is a vaporizing form of injection, the injection chamber must again have a sufficient volume to accommodate the vaporized mixture of solvent and solutes; an internal volume of ~ 1 cm^3 is probably average. Capillary columns of conventional dimensions can tolerate only restricted flows; assuming a column flow of 2 cm^3/min, one volume of carrier gas would pass through the inlet every 30 sec. As an initial target, the purge function is activated after one and one-half volumes of gas have passed through (and flushed) the inlet; in our example, the first trial would activate the purge function at 45 sec, and the results would then be evaluated in terms of solute deliveries and solvent tailing. Purge activation times slightly shorter and slightly longer should also be tried to verify the most suitable delay. Shorter purge activation times tend to decrease both solvent tailing and delivery of solutes (especially higher-boiling solutes) to the column; longer purge activation times increase both the degree of solvent tailing and delivery of solutes to the column. The existence of the purge step means that we are dealing with a contradiction in terminology: splitless injections are not "splitless"; they are splitless only in comparison to a split injection. Quantitation is usually least reliable with this injection mode.

For a given sample, the splitless injector can generally be operated at a lower temperature than would be required to introduce the same sample by split injection. In splitless injection, solutes can volatilize somewhat more slowly, because the sample band (which is continuously conducted to the column until the time of purge activation) will be subjected to a phase ratio focus.

Regardless of the injection mode, hydrogen is the preferred carrier for capillary chromatography because it yields better overall separations and shorter analysis times [1,3] (see also Chapter 5); with splitless injection, hydrogen has still another advantage. Because it is used at higher average linear carrier gas velocities, the flow through the inlet is higher and solutes are transferred from the inlet to the column more rapidly and more completely [20]. Columns coated with cross-linked (i.e., nonextractable) phases should always be used for splitless injection; normal columns suffer localized phase stripping and shortened lifetimes due to the repeated on-column solvent condensation.

3.7 Programmed Temperature Vaporizing Injector (PTV)

This specialized form of injector can be used in split, splitless, or direct mode and employs an initial inlet temperature that is below the boiling point of all components, including any solvent [23–26]. After the sample is injected, the temperature of the inlet is programmed to increase at a high rate. Each component is vaporized and moved through the split point to the column as it gains sufficient volatility from the continually increasing temperature. Hence solutes should be exposed to less thermal shock. Excellent linearities up to the C_{28} hydrocarbon have been reported. Samples can be injected neat (i.e., without solvent); where solvent is present, it vaporizes less violently and the volume of the injection chamber can be more restricted.

On-Column Injection

On-column injection as it is known today also owes its origin largely to the efforts of the Grobs (e.g., [27,28; see also 29]). This technique holds its greatest utility for two types of samples (1) those containing thermally labile solutes and

(2) those containing very high boiling solutes. In the vaporizing modes of injection (split, splitless, and "PTV"), the sample is vaporized in the heated inlet, transported to a cooler column, and finally chromatographed. Although this sequence of events safeguards the column, in that nonvolatile residues remain in the injector and the column is exposed only to materials that can be volatilized, thermally labile solutes may decompose. In on-column injection, the sample is deposited from the syringe directly on the column without undergoing vaporization. As the temperature of the column increases, solute vapor pressures rise and the rates of solute transport through the column increases (i.e., each solute experiences a decrease in k due to the decrease in c_S/c_M associated with increasing temperature). In this case, however, any nonvolatile or high-boiling solutes in the sample have also been deposited on the column. These deposits lead to severe "bleed" and baseline problems and to a rapid deterioration of column efficiency. High-temperature "baking" to remove the volatile residues in such deposits places an unnecessary stress on the column and leads to shortened column life (Chapter 6). Because they can be washed with a variety of solvents, columns coated with nonextractable cross-linked phases are preferred for on-column injection. A cautionary note seems advisable: if higher-boiling and/or nonvolatile residues are to be washed from the column, this process must be undertaken while those substances are soluble. Exposure to high temperatures for extended periods may render those deposits insoluble.

Originally, both the on-column injector and the column oven were cooled to temperatures below the boiling point of the solvent/solutes. For the analysis of higher-boiling solutes, it was then necessary to heat the oven to a point where those solutes exhibited suitable chromatographic properties. Following this, it was necessary to cool the entire assembly before the next analysis could be performed. This was inconvenient even with temperature programming and essentially impossible for higher-temperature isothermal operation. Galli *et al.* [11] suggested that "secondary cooling" offered an improvement: the column was housed in the heated oven and, during injection, cool air was discharged tangentially around the inlet end of the column. This cooled a short section of the column; following injection, the secondary cooling was discontinued, and the column assumed oven temperature.

Early results with on-column injection were sometimes disappointing; there were many reports of lowered separation efficiencies, poor quantitation, malformed peaks, and peak splitting. Knauss *et al.* [30] suggested that secondary cooling was responsible for some of these problems: the injected sample plug extended beyond the cooled zone, so that the front portion of the plug began the chromatographic process while only the latter portion was delayed by cold-trapping. This, the authors suggested, might explain "peak splitting," an occurrence that was all too common in this mode of injection. Many workers also noted that component resolution usually suffered in on-column injections and that this was exacerbated by secondary cooling. The secondary cooling-engendered loss in separation efficiency has been rationalized [12,31]; with split or splitless injections, band-shortening mechanisms come into play, as discussed above. With on-column injection, not only are these absent, but the use of secondary cooling requires the sample to advance over a positive temperature ramp; the c_S/c_M ratio is smaller at the band front than at the band rear, $v_{front} > v_{rear}$, and peaks become broader. Two complementary routes were quickly taken to correct this problem: (1) the inlet end of the column was housed in a nozzle that restricted diffusion of the cooling air so as to cool a longer segment of the column, and (2) the "retention gap" [32] was suggested as a means of reconcentrating bands and achieving band shortening. As it is most frequently employed, the retention gap is a section of uncoated, deactivated tubing attached to the inlet end of the column (see below). The concept is extremely useful in splitless and on-column injections, and by using large-diameter tubing for the retention gap, automated injectors can be used in on-column injection.

Jenkins [33] suggested that the extreme flexibility and low thermal mass of the fused silica column permitted an elegantly simple approach to cold on-column injections, regardless of the oven temperature: a short section of the column is retracted from the oven and immediately cools to ambient temperature; the injection is made into that cooled portion of the column, which is then reinserted into the heated oven. Abruptly "stepping" the column to the higher temperature avoids the band lengthening associated with the positive temperature ramp [10].

With any on-column (and, to a lesser degree, splitless) injection technique, long "smeared" solute bands can lead to serious complications. The susceptibility of the sample to this physical smearing is governed by a number of sometimes interrelated factors, including the size of the sample injected (larger samples accentuate the problem), the "wettability" of the column by the solutes/solvent (influenced by the deactivation treatment, the type and thickness of stationary phase coating, and the "polarities" of the solutes, solvent, and stationary phase), how long the liquid plug endures (the boiling points of the solvent and solutes, the temperature of the column at the point of injection, and the subsequent temperature profile), and the velocity of the carrier gas. Various aspects of these band-lengthening phenomena and methods of reconcentrating those bands have been considered in some detail (e.g., [4,14–17,34,35]).

Band lengthening due to solvent movement can be reduced by making the injection under conditions that encourage quiescent evaporation of the solvent immediately beyond the point of injection, leaving the solutes cold-trapped on the column [31]. Another approach is a "stop flow" injection, i.e., interrupting the flow of carrier gas during the injection [36–38]. These approaches can also benefit from incorporation of the retention gap concept.

3.8   Retention Gap Focusing

As discussed above, prefacing the coated separation column with a section of uncoated tubing achieves several desirable effects:

1. Nonvolatile residues tend to accumulate in the "guard column" and limit contamination of the separation column.

2. By using a precolumn of sufficient diameter, standard-sized needles can be utilized in on-column injections into small-diameter columns (which also permits automating such injections).

3. The abrupt decrease in the phase ratio effectively shortens the band beginning the chromatographic process.

4. The abrupt increase in the distribution constant also exercises a focusing effect.

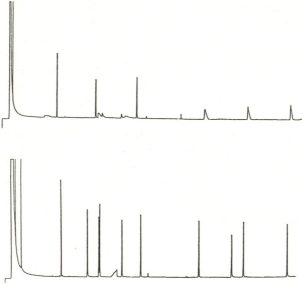

Fig. 3.5. Influence of active sites in the retention gap. Both chromatograms were obtained with cold on-column injection of the Grob test mixture through a 1-m fused silica retention gap to a 30 m × 0.32 mm fused silica column coated with DB-1. Raw fused silica tubing was used for the upper chromatogram; the tubing used in the lower chromatogram was treated with oc-tamethylcyclotetrasiloxane. The small asymmetric (leading edge, or "sloping front") peak in the lower chromatogram is not indicative of activity, but testifies to a mismatch between the polysiloxane stationary phase and this free acid solute (see Chapter 5).

On the other hand, the coating process itself has a pacifying action; very thinly coated columns often exhibit excessive activity, and uncoated precolumns can be extremely active [39]. Figure 3.5 shows severe activity contributed by a 1-m retention gap composed of raw (untreated) fused silica tubing, as contrasted with the results obtained with a section of deactivated fused silica tubing. Octamethyltetracyclosiloxane, diphenyltetramethylsiloxane, and tetraphenyldimethylsiloxane have all given excellent results when employed as vapor phase deactivating agents on fused silica tubing [39].

Selecting the Proper Injection Mode

Large-Diameter Open Tubular Columns. The larger-diameter open tubular columns that can be directly substituted for packed columns are not capillary columns and can tolerate much larger gas flow volumes. As a result, normal-size injections into these columns usually pose no special problems; they are generally attached directly to the standard packed column injector by virtue of a reducing adapter. As explored above, performance can be improved by employing injection liners specifically designed for these columns. The direct flash-vaporizing injector, in which the injected sample is rapidly vaporized and then conducted to the column (Fig. 3.4), is usually preferred for most applications. For thermally labile or high-boiling solutes, the on-column liner is usually advantageous. As mentioned above, if the latter is packed with an appropriate material and installed in an inverted position, it is possible to confine injection residues to the inlet.

Capillary Columns. With the smaller-diameter open tubular columns, gas flow volumes are much more restricted and specialized injectors are required. Several factors should be considered in deciding which injection mode is most suitable for a given sample. That decision is usually based on the nature and composition of the sample and on the goals of the investigation. Preliminary sample workup or preparation may be necessary in some cases; many such procedures are based on isolation or concentration steps that employ cold-trapping, distillation, extraction, and/or adsorption. There is often no way to avoid these steps, but it must be remembered that both quantitative and qualitative changes in sample composition can result from these treatments [10].

Direct headspace injection frequently offers the most direct and elegantly simple approach to sample introduction. A cold trap can be employed to exercise a distribution constant focus on vapor sample injections as large as 1 ml, when these are made from a macrosyringe fitted with a fused silica needle, directly into a fused silica column [6,40]. Not only does the increased inertness of this sampling system make it possible to detect some of the more reactive solutes (including sulfur- and nitrogen-containing compounds), but dilution of the sample by carrier gas is minimized by direct injection into the column* rather than into an injection chamber, whence additional carrier gas further dilutes the sample while conducting it to the column. These cold-trapped solute (or sample) bands are not stationary; unless their vapor pressures are zero, some mass of those materials enters the mobile phase and must be chromatographically transported through the column. Normally the vapor pressure is so low (and the distribution constant so small) that the steady-state concentration in the mobile phase is below the limits of detection. The amount of a solute lost to detection is usually minuscule, but the conclusion that sharp peaks testify to quantitative cold trapping is theoretically invalid. With solutes of higher vapor pressure (lower-boiling solutes or higher trapping temperatures), losses would be higher, and in cases where the mobile phase concentration of a solute is above the limits of detection, breakthrough becomes evident. Components that negotiate the trap eventually ascend a positive temperature gradient as they leave the trap and enter the uncooled column. This subjects them to a distribution constant defocus; $v_{front} > v_{rear}$, and the shapes of those peaks are badly degraded [6,40]. On-column vapor phase injections hold particular appeal where the volatiles of interest have sufficient volatility, or where their volatilities can be increased, e.g., by increasing the temperature of the sample. Such injections introduce relatively long bands to the column, and some means of achieving band focusing (as discussed earlier) should be employed prior to the chromatographic process. This subject is considered again in Chapter 8.

The inlet splitter is probably the simplest of the specialized injectors used with capillary columns, and when it is properly employed split injections can be quite satisfactory. The sample should be liquid (or in solution), and the solutes should have reasonable thermal stabilities. Further, their concentrations in the sample should be approximately 0.01–10%. If the volatilities of the solutes fall within the approximate range of the C_6–C_{19} hydrocarbons, it is usually possible to select operational conditions (temperature, flow path, internal volume, split ratio, etc.) that yield good linearity. The range of solute volatilities over which good linearity is achieved can sometimes be extended by employing programmed temperature vaporization or by varying some of the above parameters.

---

*Work in progress in the author's laboratory indicates that the standard 250 μm i.d. column can accommodate headspace injections at least as large as 10 ml by this technique. Peak heights (and sensitivity) show reasonable correlations with the size of the headspace injection. There are indications that even larger injections can be tolerated, provided the sample is injected slowly. Polythylene glycol (PEG) columns used under these conditions are rapidly destroyed by the massive amount of air. There is sufficient water in the headspace of aqueous samples to form an ice plug in the column with these larger injections. This problem can sometimes be rectified by prefacing the analytical column with 2–3 m of large-diameter (530-μm) column; trapping is executed in this larger-diameter section.

Where the solutes exist as a dilute solution of higher-boiling solutes, splitless may be the preferred injection mode. Solute volatilities should approximate those of the C_{10}–C_{26} hydrocarbons. While detectable amounts of solutes beyond this range will be delivered to the column, discrimination occurs; the severity of the discrimination varies with solute volatility and other factors. Hydrogen is almost always the preferred carrier gas in open tubular columns (see Chapter 5), but it offers some additional benefits in splitless injection. Hydrogen yields higher optimum linear velocities, which translate to larger volumetric flows; these larger flow volumes are more efficient at transporting the sample from the inlet to the column [20].

On-column injection holds its greatest utility for thermally labile solutes that tend to degrade or rearrange in the vaporizing modes of injection and for higher-boiling solutes that resist (complete) vaporization in split or splitless modes. Some on-column inlets are designed to utilize fused silica needles in sample introduction [6,40]; these offer an extremely inert system, which permits the analysis of even low levels of active solutes. Some recent developments in specialized techniques of on-column injection show promise of further extending our analytical capabilities [7].

References

1. W. Jennings, "Gas Chromatography with Glass Capillary Columns," 2nd ed. Academic Press, New York, 1980.
2. K. Grob, Jr., *J. High Res. Chromatogr.* **2**, 15 (1979).
3. K. Grob and G. Grob, *J. High Res. Chromatogr.* **2**, 109 (1979).
4. K. Grob, Jr., *J. Chromatogr.* **213**, 3 (1981).
5. K. Grob, Jr., *Anal. Proc. (London)* **19**, 233 (1982).
6. W. Jennings, *in* "Sample Introduction in Capillary Gas Chromatography" (P. Sandra, ed.), p. 23. Huethig, Heidelberg, 1985.
7. M. F. Mehran and W. Jennings, *J. Chromatogr. Sci.* **24**, 34 (1986).
8. M. F. Mehran, W. J. Cooper, M. Mehran, and W. Jennings, *J. Chromatogr. Sci.* **24**, 142 (1986).
9. J. A. Rijks, J. Drozd, and J. Novak, *in* "Advances in Chromatography" (A. Zlatkis, ed.), p. 195. Elsevier, Amsterdam, 1979.
10. W. Jennings and A. Rapp, "Sample Preparation for Gas Chromatography." Huethig, Heidelberg, 1983.
11. M. Galli, S. Trestianu, and K. Grob, Jr., *Proc. Int. Symp. Capillary Chromatogr., 3rd, 1979,* 149 (1979).
12. W. Jennings and G. Takeoka, *Chromatographia* **15**, 575 (1982).
13. W. Jennings, *J. Chromatogr. Sci.* **22**, 129 (1984).
14. K. Grob and G. Grob, *Chromatographia* **5**, 3 (1972).
15. K. Grob and G. Grob, *J. Chromatogr.* **94**, 53 (1974).
16. K. Grob and K. Grob, Jr., *J. High Res. Chromatogr.* **1**, 275 (1978).
17. R. J. Miller and W. Jennings, *J. High Res. Chromatogr.* **2**, 72 (1979).
18. R. R. Freeman, E. J. Guthrie, and L. Plotczyk (J & W Scientific, Inc.), personal communication (1985).

19. K. Grob, and G. Grob, *J. Chromatogr. Sci.* **7,** 587 (1969).
20. K. Grob, Jr. and A. Romann, *J. Chromatogr.* **214,** 118 (1981).
21. G. Schomburg, H. Behlau, R. Dielmann, F. Weeke, and H. Husmann, *J. Chromatogr.* **142,** 87 (1977).
22. G. Schomburg, H. Husmann, and R. Rittmann, *J. Chromatogr.* **204,** 85 (1981).
23. F. Poy, S. Visani, and F. Terrosi, *J. Chromatogr.* **217,** 81 (1981).
24. F. Poy, *Chromatographia* **16,** 343 (1982).
25. G. Schomburg, H. Husmann, F. Schultz, G. Teller, and M. Bender, *Proc. Int. Symp. Capillary Chromatogr., 5th, 1983,* p. 280 (1983).
26. F. Poy and L. Cobelli, *J. Chromatogr.* **279,** 689 (1983).
27. K. Grob and K. Grob, Jr., *J. Chromatogr.* **151,** 311 (1978).
28. K. Grob, *J. High Res. Chromatogr.* **1,** 263 (1978).
29. G. Schomburg, R. Diehlmann, H. Husmann, and F. Weeke, *Chromatographia* **10,** 383 (1977).
30. K. Knauss, J. Fullemann, and M. P. Turner, *J. High Res. Chromatogr.* **4,** 681 (1981).
31. R. Jenkins and W. Jennings, *J. High Res. Chromatogr.* **6,** 228 (1983).
32. K. Grob, Jr. and R. Mueller, *J. Chromatogr.* **244,** 185 (1982).
33. R. Jenkins, personal communication (1982).
34. K. Grob, Jr., *J. Chromatogr.* **251,** 235 (1982).
35. K. Grob, Jr., *J. Chromatogr.* **253,** 17 (1982).
36. F. Hougen, personal communication (1981).
37. E. Bayer and G. H. Liu, *J. Chromatogr.* **256,** 201 (1983).
38. F. Pacholec and C. F. Poole, *Chromatographia* **18,** 234 (1984).
39. W. Jennings and G. Takeoka, *in* "Neu Ulmer Gesprache 1985" (H. Jaeger, ed.). Huethig, Heidelberg, 1987 (in press).
40. G. Takeoka and W. Jennings, *J. Chromatogr. Sci.* **22,** 177 (1984).

CHAPTER 4
THE STATIONARY PHASE

4.1 General Considerations

Gas chromatographic activities were formerly dominated by packed columns, which are restricted to relatively low numbers of theoretical plates. Separation efficiences in these "low-n" columns are very much dependent on solute α values, or stationary phase selectivity [1] [see Eq. (1.23) and the accompanying discussion]. As a result, a large number of stationary phases are necessary with packed columns. Only a few of the many stationary phases that were developed over the years have been discarded; most of them are still available. Open tubular columns are capable of much higher theoretical plate numbers and can often attain superior separations with a less selective stationary phase. The analyst using open tubular columns usually regards as obsolete the vast majority of the more than 200 stationary phases listed by the supply houses catering to the packed column chromatographer. The less selective stationary phases offer several benefits: they are usually the "less polar" phases, and columns coated with those phases exhibit longer lifetimes, lower bleed rates, and shorter analysis times; the latter normally result in higher sensitivities. In most cases, the apolar column is characterized by higher theoretical plate numbers than more polar columns whose other parameters (diameter, length, d_f) are equivalent (DB-Wax is an exception to this generality). Based on individual tests of many thousands of columns, column efficiencies, in terms of theoretical plates per meter of column length, usually decrease in the following order: methylpolysiloxane = DB-Wax > DB-1301 = 1701 = 210 > DB-17 = 225.

High-quality columns prepared with apolar phases are capable of performing a large percentage of all analyses and should be used whenever possible. There are some separations that do require both high numbers of theoretical plates and greater selectivity; selected examples are discussed later.

4.2   Stationary Phase Polarity and Selectivity

The terms "selectivity" and "polarity" are often misused in gas chromatography and sometimes (incorrectly) used interchangeably. The term "polarity" can be applied to either the stationary phase or the solute(s) and implies that the substance possesses a permanent dipole. Polar stationary phases are those with appreciable concentrations of functional groups of the type —CN, —CO, —C—O, —OH, etc. [2]. Hydrocarbon-type stationary phases (e.g., squalane and the Apiezon greases) are apolar, while the polyester-type stationary phases have a much higher degree of polarity.

Both the stationary phase and the solutes play roles in "selectivity"; a stationary phase that exhibits high selectivity for one type of solute may exhibit intermediate or low selectivity for solutes with different functional groups. A generality (to which there are exceptions) is that the larger the solute K_D values are in a stationary phase, the more selectivity that stationary phase displays toward those solutes. Apolar hydrocarbon solutes would exhibit larger values of K_D in an apolar stationary phase than in a polar phase; polar solutes would exhibit larger values of K_D in a more polar stationary phase than in an apolar stationary phase. Selectivity for hydrocarbon solutes would be higher in the apolar and lower in the polar stationary phase, whereas selectivity for more polar solutes would be greater in a more polar stationary phase.

If a given solute is chromatographed under a particular set of conditions on columns coated with different stationary phases, the solute partition ratio will be largest for the stationary phase that has the higher degree of interaction with that solute. Three types of solute–stationary phase interactions are of concern in gas chromatography: dispersion, dipole, and base–acid. Most of the following discussion hinges on work directed by Hawkes [2] and contributions of Stark *et al.* [3].

Dispersion Interaction

The electrons of an atom or molecule oscillate through several different and distinct positions, each of which is characterized by a particular pattern of electrical asymmetry. Each of those short-lived electronic configurations causes the atom (or molecule) to display an overall instantaneous dipole, which leads to polarization of adjacent atoms or molecules and generates attractive forces between those atoms or molecules. These intermolecular attractions constitute the major part of the total attractive force between a hydrocarbon solute and the stationary phase, and they are significant for any dissolved solute molecules and the surrounding stationary phase solvent. There have been attempts to relate the potential for dispersion interaction to the refractive index [4]; Burns and Hawkes [2] used a modification of this approach in calculating the dispersion indices shown in Table 4.1.

Dipole Interactions

When both the stationary phase and the solute possess permanent dipole moments, the alignment of the two dipoles can result in a strong interaction between the stationary phase and the solute. In some cases the close proximity of a strong dipole in the stationary phase can generate an induced dipole in the solute (and vice versa). Such dipole interactions usually involve individual functional groups of the solute and of the stationary phase. Burns and Hawkes [2] used benzene and hexane to calculate the relative strengths of permanent dipoles in the stationary phase. They reasoned that the difference in the adjusted retentions of hexane and benzene on a given stationary phase could be attributed to an induced dipole (benzene)–permanent dipole (stationary phase) interaction, provided dispersion effects were taken into account. "Polarity" was earlier defined as the presence of a permanent dipole; hence the magnitude of the "dipole index value" in Table 4.1 can be considered a measure of the relative polarity of a given stationary phase.

Base–Acid Interactions

With hydroxyl-containing solutes that exhibit similar dipole and dispersion potentials, the formation of hydrogen bonds with the stationary phase is an important separation mechanism. The ability of a stationary phase to participate in hydrogen bonding is usually a measure of its Bronsted basicity; the hydroxyl group can, however, also behave as a base and is retained by Lewis and Bronsted acids. Because relatively few of the more commonly used stationary phases

TABLE 4.1

Stationary Phase Interaction Indices[a]

Stationary phase	Dispersion index	Dipole index	Basicity index	Acidity index
Dimethyl silicone	9	0	0	0
Phenylmethyl silicone	11.6	0	0	1
3-Cyanopropyl silicone	10.4	11	3	0
Trifluoropropyl silicone	8.6	3	0	1
Polyethylene glycol	8.6	8	4	0

[a]Data from Burns and Hawkes [2].

exhibit acidity, the increased retention of an alcohol (as compared to another solute exhibiting similar dispersion and dipole interactions) has been used as an index of stationary phase basicity [2]. These values are also shown in Table 4.1.

Gas chromatography is essentially a volatility phenomenon; the rate at which a given solute traverses the column varies directly with the average velocity of the mobile phase and indirectly with the proportion of the time the molecules of that solute spend in the stationary phase:

$$v = u/(k + 1) \qquad (4.1)$$

Both solute partition ratios k and solute vapor pressures are governed largely by (1) the temperature and (2) the degree of interaction between the solute and the stationary phase. When dispersive forces are the dominant form of solute–stationary phase interaction, both solute elution orders and the relative degree of solute separation can be predicted on the basis of solute boiling points, because these then dictate the solute vapor pressures.

The "retentiveness" of a stationary phase for a given solute under a particular set of conditions is evidenced by the partition ratio k; the more retentive the phase (toward that solute), the larger the k. The ratio of two solute partition ratios (k_2/k_1) determines their relative retention, and (assuming constant peak width) the larger the relative retention of two solutes, the greater their resolution [Eq. (1.23)].

There are several disadvantages to increased stationary phase selectivity. A larger K_D usually translates to lower sensitivity; the larger K_D, the lower the solute concentration in mobile phase and the longer the time required to transport that solute from the end of the column to the detector (at constant carrier flow). This results in broader and less intense peaks. From Eq. (1.14), as K_D increases the solute partition ratio must also increase, and longer analysis times will be required. These effects can be countered by increasing the column temperature at the time of detection (which would lower the K_D); for higher-molecular-weight solutes or those engaging in a higher degree of solute–stationary phase interaction, the higher required temperature may place increased stress on the stationary phase and lead to shorter column life.

There have been several attempts to estabish sytematic rationales for stationary phase selection (e.g., Rohrschneider or McReynolds constants), but the process is in most cases still based on empirical judgments. Where improved solute separation is the goal, a stationary phase which would be expected to engage in a greater degree of interaction with those particular solutes is employed. On the other hand, a more inert phase would be used to encourage the elution of extremely high-boiling solutes or those with more reactive functional groups. It is also possible to select stationary phase mixtures or to synthesize new stationary phases that optimize the relative retentions of a given group of solutes to yield optimized separations. This latter approach is considered in later chapters.

4.3   Polysiloxane Stationary Phases: General Comments

The polysiloxane phases have been the subject of an updating by Blomberg [1] and a thorough review by Haken [5]. They are normally viewed as constituting the most "abuse-tolerant" group of stationary phases. However, the nature of their substituent groups and their homogeneity and purity strongly influence behavior under stress; they are not indestructible. An improperly prepared glass or fused silica surface can cause phase deterioration, as can the free hydroxyl groups that are normally present in the polymers themselves. In the better stationary phase preparations, free hydroxyls have been converted to inactive groups, a process which for the terminal hydroxyls is termed "end capping." Even minute traces of the catalysts used in the preparation of the polysiloxane polymers will lead to phase deterioration; all of these problems are exacerbated by higher temperatures (see Chapter 10). High-quality columns demand stationary phases of the highest purity, homogeneity, and batch-to-batch reproducibility.

The viscosity of the methylpolysiloxanes is influenced only slightly by temperature. This has been attributed to their structure; in the absence of dilutents such as solvents, a linear polysiloxane molecule forms a coiled helix in which the siloxane bonds are shielded by orientation toward the axis and from which the alkyl groups project outward [6].

With the methylsilicones, a temperature increase brings about two opposing effects: (1) it tends to increase the mean intermolecular distance, and (2) it expands the helices and thus tends to decrease the distance. Hence the tendency to increase the mean intermolecular distance is countered by the expansion of the helix, and temperature has little effect on the viscosity of the methylsilicones. The substitution of functional groups such as phenyl or cyanopropyl (see below) disrupts the helical structure, and the viscosities of these silicones are more temperature-sensitive.

The methylsilicone polymers have several properties that are desirable in stationary phases; they can be thermally stable, viscosity is affected only slightly by temperature, and they have good wetting properties. In consequence, the majority of stationary phases employed today are polysiloxane polymers, based on an oxygen–silicon lattice that can be represented as

$$
\begin{array}{ccc}
\text{R} & \text{R} & \text{R} \\
| & | & | \\
\text{R—Si-O[— Si—O]}_n\text{—Si—R} \\
| & | & | \\
\text{R} & \text{R} & \text{R}
\end{array}
$$

4.4   Dimethyl Polysiloxane Stationary Phases

In the dimethyl polysiloxanes, the R positions are occupied by methyl groups. To impart greater "functionality," other groups may be substituted for some of

TABLE 4.2

Composition of Selected Stationary Phases

Phase	CH$_3^a$ (%)	Phenyl (%)	CNPr^b (%)	Other
SE-30, OV-101, DB-1	100	0	0	0
SE-54, DB-5	94	5	0	1% vinyl
OV-17, DB-17	50	50	0	0
OV-210, DB-210	50	0	0	50% TFP^c
OV-225, DB-225	50	25	25	0
DB-1301	94	3	3	0
OV-1701, DB-1701	88	6	6	0
SP-2330, DB-2330	25	0	75	0

^{a}In the bonded (DB) polysiloxane phases, the percent CH$_3$ figure is nominal; a few of the bonds indicated as methyl exist as cross-links or surface bonds.
^{b}Cyanopropyl.
^{c}Trifluoropropyl.

the methyl groups. Table 4.2 lists the approximate compositions of some of the more popular silicone phases.

The silicone oils (e.g., SF-96, OV-101, SP-2100) are linear dimethyl polysiloxanes that are normally fluid; the silicone gums (e.g., OV-1, SE-30) are also linear dimethyl polysiloxanes, but of higher molecular weight; the term "silicone rubber" usually but not always denotes a cross-linked (and sometimes surface-bonded; see below) silicone gum (e.g., DB-1, Ultra-1) [5]. The wetting characteristics of these methyl silicones are such that they form films of the highest uniformity on glass and fused silica; columns produced with these stationary phases not infrequently exhibit the ultimate in theoretical efficiencies.

The major limitation of the methyl silicone stationary phases is their lack of functionality. The interaction between solutes and a dimethyl silicone phase is limited largely to dispersion forces; because solute elution order is based on solute vapor pressures and extradispersion forces (e.g., hydrogen bonding, dipole–dipole) play little or no role, solute elutions occur in the order of solute boiling points. Solutes that cannot be sufficiently differentiated on the basis of their dispersive interactions require the use of stationary phases with which they can engage in other types of interactions.

4.5   Other Silicone Stationary Phases

On those occasions when stationary phases of greater selectivity are desirable, the incorporation of phenyl, vinyl, cyanopropyl, or trifluoropropyl groups in place of some of the methyl substituents on the polysiloxane chain produces polymers that not only can engage in dispersive interactions, but also are capable of different degrees of dipole and/or acid–base interactions; these will display increased selectivity toward particular solutes. Such substitutions will also affect the thermal stability of the polymer [7]. Referring back to the schematic structure just preceding Section 4.4, substitution of an electronegative group for R would decrease the strength of the Si–C bond and increase that of the Si–O–Si bond; conversely, substitution of an electron donor would strengthen the Si–C bond and weaken the Si–O–Si bond. This would lead one to predict that substituting phenyl for much of the methyl would yield a higher-temperature polysiloxane phase [7]. Although it is true that the thermal stability of the polymer itself is improved, that greater stability is often of only limited utility to the chromatographer. The surface of the fused silica tubing requires other special treatments to encourage wetting by this high-phenyl polymer, and the thermal limits of those treatments may now dictate the upper temperature limit of the column, as reflected in Table 4.2.

Cyano groups are also strongly electron-attracting, and their substitution on the polysiloxane chain results in weakening the Si–C bond. This is especially true if the cyano group is in the α position; a γ-substituted cyanosilicone reportedly has essentially the same oxidative thermal stability as a methyl silicone [1,7].

4.6   Bonded, Cross-Linked, and/or Immobilized Stationary Phases

"Bonded" phases, as the term is usually used in open tubular columns, resulted from the commercial discovery that conventional glass capillary columns coated with SE-54 by a high-temperature, high-pressure process [8] exhibited greatly extended lifetimes. It was deduced that the vinyl moiety of the SE-54 stationary phase reacted under these very stringent conditions to produce a cross-linked stationary phase. That discovery was exploited by incorporating into other prepolymeric oligimers a small amount of vinyl; both high temperatures and peroxide additions were used to initiate free radicals, and "bonded phase" columns became commercially available [9]. Grob [10] reported his first findings on "immobilization" of the OV-61 stationary phase at almost the same time, indicating that he had been working independently along the same lines. That publication was soon followed by the "immobilization" of OV-1701 [11] and a "cross-linked surface-bonded" OV-1701 [12]. From this point in time, developments occurred rapidly; while the practicing analyst continues to reap benefits from these activities, their consideration is beyond the scope of this book. The interested reader is referred to further publications of K. Grob, G. Grob, and K. Grob, Jr., and to those from groups headed by L. Blomberg, G. Schomburg, M. L. Lee, and S. Lipsky, as well as many others.

Initially, the term "bonded" elicited considerable controversy over its meaning and usage, resulting in some degree of semantic confusion; some equated the wetting forces holding deposited films with covalently bonded films; others debated whether these modified stationary phases were surface bonded or cross-linked, and there were suggestions that the distinction was perhaps unimportant and that all such stationary phases should be termed "immobilized." There is, in fact, a difference; some phases are merely cross-linked, others are both cross-linked and bonded to the surface by means of covalent bonds [9,12]. Under normal circumstances the distinction is probably unimportant to the average gas chromatographer. It is of greater importance to those using splitless and on-column injection, and it can become painfully clear to those using such columns for open tubular liquid chromatography or for supercritical fluid chromatography. Columns that are coated with stationary phases that are both cross-linked and surface-bonded exhibit longer lifetimes in these applications [13].

Current procedures for column preparation usually involve pretreatments of the interior surface of the fused silica tubing; some of the surface preparation steps are designed to render the fused silica surface silanols amenable to stationary phase bonding, are considered proprietary, and fall outside the scope of this book. Peroxides may be added to the coating solution, and static coating procedures [14–16] are generally employed to deposit a film of vinyl-containing stationary phase oligomers on the interior wall of the tubing. Heating causes peroxide decomposition, which yields free radicals and initiates cross-linking (and, if the surface has been properly prepared, surface bonding) when the column is heated. The preferred peroxide is generally dicumyl peroxide, which decomposes to form volatile products that are dissipated during the heating step [9]. The use of ozone to initiate cross-linking in both nonpolar and medium-polar silicone stationary phases has been suggested [17], and attention has also been directed to the use of azo compounds [18,19]. A methyl-2-phenylethylpolysiloxane was reported to undergo cross-linking without addition of a free radical initiator [20]; in this respect, it is similar to the SE-54 which really started all of this activity. Ionizing radiation has also been used as a free-radical initiator for the polysiloxanes [21–24], but such treatments can have detrimental effects on the flexibility and integrity of the protective outer polyimide coating; imperfections in that coating can lead to column breakage [25].

There have also been several recent publications on the synthesis of other polysiloxane stationary phases [26–29] and on methods of characterizing stationary phases by liquid chromatography [30], by supercritical fluid chromatography [31], and by gas chromatography of their hydrolytic products [32].

4.7   Polyethylene Glycol Stationary Phases

Probably because its higher average molecular weight permits its use at higher temperatures, Carbowax 20M has been the polyethylene glycol most widely used in gas chromatography. Structurally, it can be represented as

$$HO—CH_2—CH_2—[O—CH_2—CH_2]_n—OH$$

While the average molecular weight is 20,000, the range is not specified and wide batch-to-batch variations occur. One of its major disadvantages is oxygen lability, especially at higher temperatures. In common with the silicone phases, even traces of oxygen are detrimental at higher temperatures, but with the polyethylene glycol phases, even with less oxygen and lower temperatures, the penalty is higher. Reducing the oxygen concentration in the carrier gas can help prolong the lifetime of columns coated with polyethylene glycol phases [33]. The decomposition products include acetaldehyde and acetic acid [34].

Other disadvantages for which Carbowax 20M is faulted include its solubility in water and low-molecular-weight alcohols and its relatively high low-temperature limit: it solidifies to a waxy solid at 50–60°C. The first of these poses limitations on the types of samples that can be used, especially with splitless and on-column modes of injection.

The high low-temperature limit can also be troublesome. In the preceding chapter, the influence of solute diffusivity in the stationary phase was explored. With stationary phases of normal diffusivity ($D_S \geq 10^{-6}$ cm^2/sec [35,36]), this term has little influence in open tubular columns until d_f exceeds about 0.4 μm. At larger values of d_f or smaller values of D_S, chromatography suffers. If a stationary phase that is normally liquid is operated under conditions where it becomes a solid, D_S increases greatly. The lower operating temperature limit of a stationary phase is normally dictated by this provision (Fig. 4.1 [37]).

There has also been progress in bonding polyethylene glycol stationary phases. Methods that have been reported include coating leached Pyrex glass with Carbowax 20M and then recoating with a second solution containing the same stationary phase plus dicumyl peroxide [38]. The use of dicumyl peroxide to bond Carbowax 20M to an adsorbed layer of graphitzed carbon has been

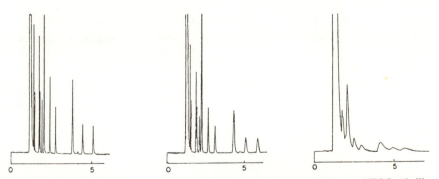

Fig. 4.1. Isothermal chromatograms of a polarity test mixture on a Carbowax 20M fused silica column at several temperatures. The anomolous results at lower temperatures reflect the fact that the stationary phase is no longer liquid. Temperatures (left to right): 55, 50, and 45°C. (Adapted from the *Journal of Chromatographic Science*, volume 22, 1984, p. 129, with permission of the copyright holder, Preston Publications, Inc.)

described [39]; the columns reportedly tolerate "several" water injections, but the phase is extractable with dichloromethane. Some preliminary results have also been reported by Morandi *et al.* [40].

Although several commercial companies offer columns coated with bonded forms of polyethylene glycol, no details of the procedures used in their manufacture have been released. It is apparent that different methods are employed, because columns from different suppliers differ in minimum and maximum operating temperature limits and in resistance to extractability with solvents such as water or methanol. In these polar stationary phases, the lower operating temperature limit is often more critical than the upper operating temperature limit; those who find themselves limited by the upper temperature limit are usually using a stationary phase that is too retentive (polar) for the separation in question and are forced to employ higher temperatures to counter that excess polarity. For analysts concerned with the separation of low-molecular-weight polar solutes, the minimum operating temperature becomes crucial; if the temperature must be elevated in order to keep the stationary phase liquefied, separation of lower-molecular-weight solutes is severely handicapped.

4.8   Enantiomer Separations

In many cases, the biological effects of two members of an enactiomorphic pair are entirely different. Enantiomers often exhibit differences in degree of interaction with a biological receptor, in route and/or mechanism of biological transport, and in the manner in which they are metabolized. Enantiomer differentiation is important in the drug and pharmaceutical, insect pheromone, and food and flavor chemistry fields, and indeed in any field where the materials in question interact with biological systems. In addition, enantiomer separations have applications in forensic analysis, in dating of some fossil materials, and (in some cases) in the detection of synthetics added to a "natural" product.

A solute chromatographs at a velocity that can be represented as (Chapter 3)

$$v = u/(k + 1)$$

Since $K_D = \beta k$, k is a function of K_D, which is affected not only by the normal physical equilibria (e.g., dispersion forces) but also by any interplay between the solute and the stationary phase (or, for that matter, and to our occasional distress, the solid support). It is really this chemical interplay that determines the "selectivity" of a stationary phase for a given solute.

The distinction between diastereomers and enantiomers is important in the following discussions. The former are not enantiomers, and their chromatographic separation usually poses no special problem because they differ structurally, and their physical and/or chemical properties in the chromatographic process must be different [41]. Enantiomers are identical in all properties except chirality and must exhibit identical physical properties in any isotropic medium. As emphasized by Lochmueller [42], because the solute partition ratio is a function of the solute vapor pressure and the activity coefficient, and because the relative retention of two solutes is the ratio of their partition ratios:

$$\alpha = k_2/k_1 = y_1 p_1^0 / y_2 p_2^0 \qquad (4.2)$$

where y represents the activity coefficient and p^0 the vapor pressure of the solute in question. In the case of diastereomers, vapor pressures differ and separation is possible on achiral stationary phases. The vapor pressures of enantiomers are identical, and to achieve their separation, they must be chromatographed under conditions where their activity coefficients differ; this can occur only in a chiral stationary phase.

For this reason, the stationary phases considered to this point are generally unable to distinguish between the enantiomers of optically active compounds. There are two ways in which such compounds can be separated: (1) if one enantiomer of an optically active derivatizing reagent can be obtained in a high degree of purity, the resultant diastereomers (which are no longer enantiomers) can be separated on a standard stationary phase [43,44]; (2) by direct separation of the enantiomers or the enantiomeric derivatives on a chiral stationary phase [45,46].

The first route has been faulted for several disadvantages: (1) the deprivatizing reagent must be optically active, which greatly limits the choices, (2) the reagent must be optically pure, (3) systematic errors arise due to differences in the reaction kinetics of the two enantiomers, and (4) racemization may occur at one of the asymmetric carbons during derivatization.

A chiral phase, on the other hand, permits the use of any common derivatizing reagent, followed by direct separation of the reactants. The hydrogen-bonded associations between the chiral phase and the two enantiomers differ widely [46], and separation is achieved. A large number of chiral phases have been synthesized and investigated; most of those efforts have been well reviewed (e.g., [47,48]). Most such stationary phases can be classified as dipeptide phases, derivatized amino acid phases, or polymeric chiral phases. The maximum operating temperatures are often low, due to the vapor pressures of these materials or to their tendency to racemize and lose chiral specificity at higher temperatures. The temperature restriction often leads to long analysis times, low sensitivity, and inability to elute some materials. The phase which has gained the most attention to this time is probably Chirasil-Val, whose useful temperature range has been listed as 70–240°C [48]. Some recent work on bonded polymeric chiral phases [49] opens still another possibility. If these are bonded sufficiently well that they can be employed in open tubular supercritical fluid chromatography, it may soon be possible to resolve even higher-boiling enantiomers with high resolution.

A cautionary note related to enantiomer separations seems advisable. Because many of the solutes for which enantiomer differentiation is important have limited thermal stability, splitless and on-column modes of injection are often employed. Either of these injection modes can generate split peaks, which have in the past led to erroneous conclusions. It is usually worthwhile to establish that a similar injection of a single compound of similar functionality, polarity, and boiling point produces a single peak under the same conditions.

4.9   Other Special-Selectivity Stationary Phases

Stationary phases capable of certain specific types of interactions with specific types of solutes are occasionally used to perform specialized separations. The separation of enantiomers by chiral stationary phases described above is one such example that has significance to most biologically oriented analysts. Some other selective stationary phases have been reported, but their applicability is less general, and our treatment of these will in most cases be limited to a few key references.

Unsaturated and aromatic hydrocarbons have been separated on stationary phases that exercised their selectivity partially by formation of a pi-charge complex. Typical of these electron donor stationary phases are dialkyltetrachlorophthalates and 2,4,7-trinitro-9-fluorenone [50–55].

Liquid crystal stationary phases have been used for the separation of some isomeric solutes having structural rigidity, such as certain of the polychlorinated biphenyls, substituted benzenes, polynuclear aromatic hydrocarbons, and steroids [56–58]. Within a specific temperature range, the properties of a liquid crystal phase are intermediate between those of a liquid and those of a crystalline solid. Mechanically, it more nearly resembles a liquid, but a liquid with a high degree of orderliness. As a result, it still has some of the anisotropic properties of the solid. A thermotropic liquid crystal is one in which the liquid crystal state is generated above the melting point of the solid and exists through a discrete range up to some "clearing temperature," where it becomes an isotropic liquid.

The thermotropic liquid crystals are classified into nematic, cholesteric, and smectic types, depending on the liquid crystalline structure. Smectic phases have highly ordered layers of parallel rodlike molecules; the interlayer dimension is determined by the length of those molecules. Nematic phases maintain the parallel orientation, but do not exhibit the layered structure; the movement of liquid molecules is restricted only by the parallel configuration. The cholesteric phase requires that the liquid crystal have a chiral center, resulting in a twisted nematic structure. Both the smectic and cholesteric types generally exhibit the liquid crystal structure only over a very narrow range of temperatures; the nematic type usually has a broader range of operating temperature and is dominant among the liquid crystals that have been used as stationary phases [56–58].

The retention characteristics of liquid crystal columns have been reported to vary unpredictably [59,60], and they are usually characterized by very low theoretical plate numbers and poor separation efficiencies [59]. Laub *et al.* [61] reported that more efficient columns containing the liquid crystal N,N'-bis(p-butoxybenzylidene)-a,a'-bi-p-toluidine (BBBT) could be prepared by dissolving the liquid crystal in the 5% phenyl polysiloxane SE-52. Direct comparisons of their results are precluded by the massive shift of the partition ratio of the test solute, triphenylene (6.92 in SE-52; 29.4 in BBBT), but the column efficiencies seemed low. Crude comparisons indicate that the theoretical plate numbers ranged from approximately 70% of that theoretically possible for the pure SE-52, to 16% for the pure BBBT, to 43% for 80% BBBT in SE-52. Markides *et al.* [62] used aliphatic hydrocarbons as flexible spacers to couple mesomorphic moieties to a polysiloxane backbone. The result, a smectic biphenylcarboxylate ester–polysiloxane polymer, was capable of undergoing cross-linking and reportedly produced highly efficient columns whose smectic properties endured from 118 to 300°C [63].

4.10   Gas–Solid Adsorption Columns

Porous layer open tubular (PLOT) and support-coated open tubular (SCOT) columns contain a layer of adsorptive material affixed to the column wall; with adsorptive-type supports, the separation mechanism is gas–solid adsorption chromatography rather than the gas–liquid partition chromatography with which we have dealt so far. Coated with materials such as Al_2O_3, these work very well for the analysis of light hydrocarbons at above-ambient temperature. Examples of such an analysis, as well as some unique separations requiring PLOT columns coated with silica gel or certain of the porous polymers, are discussed in Chapter 9.

References

1. L. Blomberg, *J. High Res. Chromatogr.* **5**, 520 (1982).
2. W. Burns and S. J. Hawkes, *J. Chromatogr. Sci.* **15**, 185 (1977).
3. T. J. Stark, P. A. Larson, and R. D. Dandeneau, *Proc. Int. Symp. Capillary Chromatogr., 5th, 1983*, p. 65 (1983).
4. R. A. Keller, B. L. Karger, and L. R. Snyder, *in* "Gas Chromatography 1970" (R. Stock and S. G. Perry, eds.), p. 125. Inst. Petrol., London, 1971.
5. J. K. Haken, *J. Chromatogr.* **300**, 1 (1984).
6. H. W. Fox, P. W. Taylor, and W. A. Zisman, *Ind. Eng. Chem.* **39**, 1401 (1947).
7. W. Noll, "Chemistry and Technology of Silicones." Academic Press, New York, 1968.
8. W. Jennings, K. Yabumoto, and R. H. Wohleb, *J. Chromatogr. Sci.* **12**, 344 (1974).
9. R. Jenkins (J & W Scientific, Inc.), personal communication (1979).
10. K. Grob, G. Grob, and K. Grob, Jr., *J. Chromatogr.* **213**, 211 (1981).
11. K. Grob and G. Grob, *J. High Res. Chromatogr.* **5**, 13 (1982).

12. K. Grob and G. Grob, *J. High Res. Chromatogr.* **6,** 153 (1983).
13. R. Houcks (Suprex, Inc.), personal communication (1985).
14. J. Bouche and M. Verzele, *J. Gas Chromatogr.* **6,** 501 (1968).
15. M. Giabbai, M. Schoults, and W. Bertsch, *J. High Res. Chromatogr.* **1,** 277 (1978).
16. K. Grob, *J. High Res. Chromatogr.* **1,** 93 (1978).
17. J. Buijten, L. Blomberg, S. Hoffmann, K. Markides, and T. Wannman, *J. Chromatogr.* **283,** 341 (1984); **289,** 143 (1984).
18. M. L. Lee, P. A. Peaden, and B. W. Wright, *Pittsburgh Conf. on Analy. Chem. and Appl. Spectros., 1982,* Abstract 298 (1982).
19. S. R. Springston, K. Melda, and M. V. Novotny, *J. Chromatogr.* **267,** 395 (1983).
20. J. S. Bradshaw, S. J. Crowley, C. W. Harper, and M. L. Lee, *J. High Res. Chromatogr.* **7,** 89 (1984).
21. G. Schomburg, H. Husmann, S. Ruthe, and M. Herraiz, *Chromatographia* **15,** 599 (1982).
22. W. Bertsch, V. Pretorius, M. Pearce, J. C. Thompson, and N. G. Schnautz, *J. High Res. Chromatogr.* **5,** 432 (1982).
23. Gy. Vigh and O. Etler, *J. High Res. Chromatogr.* **7,** 620 (1984).
24. O. Etler and Gy. Vigh, *J. High Res. Chromatogr.* **7,** 700 (1984).
25. J. A. Huball, P. DiMauro, E. F. Barry, and E. Chabot, *J. High Res. Chromatogr.* **6,** 241 (1983).
26. J. C. Kuei, J. I. Shelton, L. W. Castle, R. C. Kong, B. E. Richter, J. S. Bradshaw, and M. L. Lee, *HRC CC, J. High Resolut. Chromatogr. Chromatogr. Commun.* **7,** 13 (1984).
27. M. Ahnoff and L. Johansson, *Chromatographia* **19,** 151 (1984).
28. J. S. Bradshaw, N. W. Adams, B. J. Tarbet, C. M. Schregenberger, B. S. Johnson, M. B. Andrus, K. E. Markides, and M. L. Lee, *Proc. Int. Symp. Capillary Chromatogr., 6th, 1985,* p. 132 (1985).
29. M. W. Ogden and H. M. McNair, *Proc. Int. Symp. Capillary Chromatogr., 6th, 1985,* p. 149 (1985).
30. J. F. K. Huber and G. Reich, *J. Chromatogr.* **294,** 15 (1984).
31. L. F. Hanneman, G. R. Sumption, and J. D. Schwake, *Proc. Int. Symp. Capillary Chromatogr., 6th, 1985,* p. 144 (1985).
32. I. Temmerman, P. Sandra, and M. Verzele, *J. High Res. Chromatogr.* **8,** 513 (1985).
33. J. R. Conder, N. A. Fruitwala, and M. K. Shingari, *J. Chromatogr.* **269,** 171 (1983).
34. J. Debraurwere and M. Verzele, *J. Chromatogr. Sci.* **14,** 296 (1976).
35. L. Butler and S. Hawkes, *J. Chromatogr. Sci.* **10,** 518 (1972).
36. J. M. Kong and S. J. Hawkes, *J. Chromatogr. Sci.* **14,** 279 (1976).
37. G. Takeoka and W. Jennings, *J. Chromatogr. Sci.* **22,** 177 (1984).
38. V. Martinez de la Gandara, J. Sanz, and I. Martinez-Castro, *J. High Res. Chromatogr.* **7,** 44 (1984).
39. M. V. Russo, G. C. Goretti, and A. Liberti, *J. High Res. Chromatogr.* **8,** 535 (1985).
40. F. Morandi, D. Andreazza, and L. Motta, *Proc. Int. Symp. Capillary Chromatogr., 6th, 1985,* p. 103 (1985).
41. C. J. W. Brooks, M. T. Gilbert, and J. D. Gilbert, *Anal. Chem.* **45,** 896 (1973).
42. C. H. Lochmueller, *Sep. Purif. Methods* **8,** 21 (1979).
43. W. A. Konig, W. Rahn, and J. Eyem, *J. Chromatogr.* **133,** 141 (1977).
44. E. Gil-Av, B. Feibush, and R. Charles-Sigler, *in* "Gas Chromatography" (A. B. Littlewood, ed.), p. 227. Inst. Pet., London, 1966.
45. W. Parr, J. Pleterski, C. Yang, and E. Bayer, *J. Chromatogr. Sci.* **9,** 141 (1971).
46. H. Frank, G. J. Nicholson, and E. Bayer, *J. Chromatogr.* **146,** 197 (1978).
47. W. A. Koenig, *J. High Res. Chromatogr.* **5,** 588 (1982).
48. V. Schurig, *Angew. Chem., Int. Ed. Engl.* **23,** 747 (1984).

49. G. Schomburg, I. Benecke, and G. Severin, *Proc. Int. Symp. Capillary Chromatogr., 6th, 1985,* p. 104 (1985).
50. D. L. Meen, F. Morris, and J. H. Purnell, *J. Chromatogr. Sci.* **9,** 281 (1971).
51. M. Ryba, *Chromatographia* **5,** 23 (1972).
52. S. H. Langer, *Anal. Chem.* **44,** 1915 (1972).
53. J. H. Purnell and O. P. Srivastava, *Anal. Chem.* **45,** 1111 (1973).
54. R. J. Laub and P. L. Pecsok, *J. Chromatogr.* **113,** 47 (1975).
55. C. L. de Ligny, *Adv. Chromatogr.* **14,** 265 (1976).
56. H. Kelker, *Adv. Liq. Cryst.* **3,** 237 (1978).
57. G. M. Janini, *Adv. Chromatogr.* **17,** 231 (1979).
58. Z. Witkiewicz, *J. Chromatogr.* **251,** 311 (1982).
59. J. Szulk and Z. Wittkiewicz, *J. Chromatogr.* **262,** 141 (1983).
60. G. M. Janini and M. T. Ubeid, *J. Chromatogr.* **248,** 217 (1982).
61. R. J. Laub, W. L. Roberts, and C. A. Smith, *J. High Res. Chromatogr.* **3,** 355 (1980).
62. K. E. Markides, H.-C. Chang, C. M. Schregenberger, B. J. Tarbet, J. S. Bradshaw, and M. L. Lee, *J. High Res. Chromatogr.* **8,** 516 (1985).
63. K. E. Markides, M. Nishioka, B. J. Tarbet, J. S. Bradshaw, and M. L. Lee, *Anal. Chem.* **57,** 1296 (1985).

CHAPTER 5

VARIABLES IN THE GAS CHROMATOGRAPHIC PROCESS

5.1 General Considerations

A number of analytical variables influence the chromatographic results obtained for a given sample. Some are under the control of the operator up to the time of sample injection and might be termed operational parameters. These include the temperature of the analysis and (for nonisothermal runs) the program parameters, as well as the column head pressure or, more precisely, the linear velocity of the carrier gas. Other operational variables depend on the injection mode, i.e., the split ratio with split injection and the purge activation time with splitless injection. The choice of solvent can also influence the results obtained, especially with splitless and on-column injections; one aspect of this concerns the relative "polarities" of the solute, solvent, and stationary phase. In all cases, syringe and injection techniques can have a variety of effects.

There are other variables that, immediately preceding sample injection, are no longer under operator control; these normally include the choice of carrier gas, the choice of injection and detection modes, certain design features of the injector and the detector (including whether or not a "retention gap" was incorporated, the manner in which it was attached, and the method by which it was deactivated), and the column variables of length, diameter, and type and thickness of the stationary phase coating. This latter group might better be considered as design parameters.

Some aspects of some of these variables are most logically fitted under headings considered in other sections. The suitability (for a given sample) of a particular injection mode is covered primarily in Chapter 3, with some additional comments relative to specialized problems in Chapter 8; the influence of instrumental design features (e.g., excessive or unswept volumes in the flow path, suitability of the detection mode, speed and fidelity of the data-handling equipment) are also considered in Chapter 7. Factors influencing the selection of the stationary phase are considered in Chapters 4 and 6.

The present chapter is concerned primarily with the proper choice of carrier gas and the effects that variables such as carrier gas velocity, column length, column diameter, and stationary phase film thickness may be expected to have on the chromatographic results; in later chapters, this information will be used to rationalize decisions concerned with stationary phases, column selection, instrument conversion and adaptation, special analytical considerations, and even specific applications. The later portion of the present chapter will also consider how changes in these variables can affect the solute elution order.

One of the decisions most frequently faced by the practicing chromatographer is the selection of carrier gas velocity; it is the first decision to be made after column installation, and it recurs whenever any of a number of parameters are changed. This seemingly simple decision can have a profound effect on separation efficiency, analysis time, sensitivity, and even the extent to which thermally labile solutes survive the chromatographic process.

5.2   Regulation of the Gas Velocity

The carrier gas velocity can be adjusted by means of a pressure regulator that controls the column head pressure or by means of a mass flow controller. Until quite recently, most open tubular columns were pressure-regulated, while flow-regulated systems, in which the increase in viscosity is countered by an increase in pressure to maintain a constant flow, were more common on packed column instruments. Flow regulation is now occasionally used on open tubular columns. Systems that utilize injection splitters direct only a fraction of the gas flow to the column, and flow regulation would be futile.

Gas viscosities vary directly with temperature; as the temperature increases, so does the viscosity of the carrier gas, and at constant inlet pressure the linear carrier gas velocity and the flow rate decrease. Hence, in a pressure-regulated system, the carrier gas velocity is not constant through the course of a programmed run but continually decreases as the program proceeds to higher temperatures. Methods of determining the carrier gas velocity were described in Chapter 1.

5.3   The van Deemter Curve

The effects of the carrier gas velocity are explored by the van Deemter equation (see Section 1.6). Some practitioners tend to resist the consideration and interpretation of van Deemter curves, but they can be extremely useful in the qualitative and quantitative evaluation of chromatographic variables. In the vast majority of cases, the interpretation of van Deemter curves is really quite simple; with the evaluation of three parameters, all useful information has been extracted. Those parameters are

1. The magnitude of h_{min}; the smaller h_{min}, the larger the number of theoretical plates per meter of column length. In the comparison of variables, that curve exhibiting the lower h_{min} denotes greater separation efficiency, all other things being equal.

2. The magnitude of $\bar{u}_{opt}$; although most analyses should usually be performed at the OPGV (discussed below), systems generating higher values of $\bar{u}_{opt}$ indicate shorter analysis times at either $\bar{u}_{opt}$ or OPGV. Because there is less time for diffusion and dilution of the solute with carrier gas, shorter analysis times generally result in higher sensitivities, all other things being equal.

3. The "flatness" of the curve; in general, steep curves are undesirable and flat curves are preferable. The reasons for this are explored below.

The compressibility of the mobile gas phase leads to diffusivity and velocity gradients through the column. At the inlet end of the column, the gas pressure and gas density are highest and diffusivities and linear velocity are lowest, while at the outlet end of the column, pressure and density are lowest and diffusivities and linear velocity are highest.

Some interrelationships of these variables were considered in Section 1.6. The fullest optimum utilization of the stationary phase within any, section of the column requires that the linear velocity of the mobile phase be in balance with the rate of solute diffusion in that mobile phase. In areas where diffusion is slow (inlet end of the column), solute molecules require more time to diffuse transversely across the column and utilize the stationary phase in a given length of column as fully as possible. In areas where transverse diffusion is more rapid (outlet end of the column), the same solute requires less time for equivalent utilization of the stationary phase in a comparable length of column. At the inlet end of the column, where both diffusion and the linear velocity of the mobile phase are lower, the lower diffusivity engendered by the higher gas density inhibits lengthening by longitudinal diffusion; this would otherwise (because of the longer exposure time) lead to inordinate lengthening of solute bands. At the outlet end, the higher linear velocities are commensurate with the higher rates of diffusion; inordinate band lengthening under these conditions of more rapid longitudinal diffusion is countered by the shorter exposure time, due to the higher linear velocity. It has been suggested that the opposing effects of the velocity and diffusivity gradients make them self-compensating; i.e., if the mobile phase velocity is optimized at one point in the column, it is optimized throughout the column [1,2]. There are some indications this may not be entirely true [3]; these relationships are considered further in Chapter 8.

5.4 Optimum Practical Gas Velocity

The van Deemter curve shown in Fig. 5.1 is a "theoretical" curve; it ignores the fact that the average linear gas velocity can be increased only by changing the pressure drop through the column, which will change the velocity and diffusivity gradients that are engendered by the compressibility of the gaseous mobile phase.

In examining Fig. 5.1, note that h_{min} amounts to 0.2 mm and occurs at a $\bar{u}_{opt}$ of 54 cm/sec. Assuming a column length of 30 m, a solute partition ratio of 5.0, and operation at $\bar{u}_{opt}$, then

$$t_M = (3000 \text{ cm})/(54 \text{ cm/sec}) = 55.6 \text{ sec}$$

$$t_R = (1 + k)t_M = 334 \text{ sec}$$

$$n = L/h = 30{,}000 \text{ mm}/0.2 \text{ mm} = 150{,}000 \text{ theoretical plates}$$

For the solute under consideration, the 150,000 theoretical plates were realized in an analysis time of 334 sec; i.e., the system generated 449 theoretical plates per second under these conditions.

If the analysis had been conducted instead at twice $\bar{u}_{opt}$ (i.e., at $\bar{u} = 111.2$ cm/sec, Fig. 5.2), the following would apply:

$$h = 0.33 \text{ mm}$$

$$t_M = 27 \text{ sec}$$

$$t_R = 162 \text{ sec}$$

$$n = 90{,}909 \text{ theoretical plates}$$

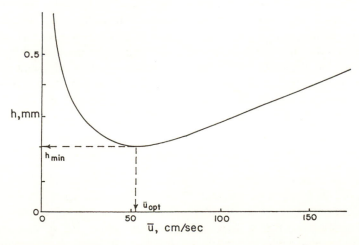

Fig. 5.1. Theoretical van Deemter curve exhibiting an h_{min} of 0.2 mm at an optimum average linear carrier gas velocity ($\bar{u}_{opt}$) of 54 cm/sec.

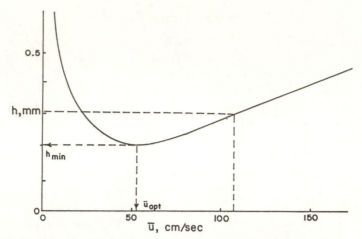

Fig. 5.2. Evaluating the effects of operating the system depicted in Fig. 5.1 at an average linear carrier gas velocity of twice the optimum value; the magnitude of h increases from 0.2 to 0.33 mm, while isothermal analysis times are halved.

In moving from $\bar{u}_{opt}$ to $2 \times \bar{u}_{opt}$, the total number of theoretical plates declined from 150,000 to 90,909 and the analysis time decreased from 278 to 162 sec; the higher velocity generated 561 theoretical plates per second. The highest number of total plates is obtained at $\bar{u}_{opt}$, but in terms of theoretical plates per unit time, velocities higher than $\bar{u}_{opt}$ would appear to be more efficient.

The optimum practical gas velocity (OPGV) has been defined as the point where the van Deemter curve becomes linear [4]; attainment of linearity, however, occurs only with theoretical curves, because the gas velocity can be varied only by changing the pressure drop through the column. Hence D_M, which is different at different points in the column and which affects both longitudinal and transverse diffusion, changes at any one point in the column as the gas velocity is varied. Experimental van Deemter curves do not exhibit linearity, but continue to curve upward. The OPGV is consequently better defined as the point where the n/t_R ratio is maximized. Because n varies with k (Section 5.7), the OPGV must also vary with k. The practical aspects of both van Deemter and OPGV curves are considered in greater detail in Chapter 6.

5.5   Computer-Generated Curves

Both van Deemter and OPGV curves can be extremely useful in exploring the effects of chromatographic variables. The generation of a series of experimental curves, however, would be formidable at the least, and in some cases impossible. To explore a variable such as the stationary phase film thickness (d_f), all other parameters, including not only column length, column diameter, solute, temperature, and stationary phase characteristics but also the efficiencies of injection, column deactivation, and column coating, would have to be precisely the same; certainly the last is rarely under precise control. However, computerized analysis of the various parameters is possible and should provide useful information.

This route was utilized by Ingraham *et al.*, employing an Apple II plus, disk operating systems, and language card, interfaced to an IDS 460G impact printer [5]. The basic van Deemter equation [6]; as modified by Giddings *et al.* [7], Giddings [1,8], Sternberg [2], and Cramers *et al.* [9], was subjected to a slight further modification. Equation 5 from the latter authors was substituted into their equation 8, resulting in the relationship

$$h = \frac{2D_M f_1 f_2}{\bar{u}} + \frac{11k^2 + 6k + 1}{96(1 + k)^2} \frac{d^2 f_1 \bar{u}}{D_M f_2} + \frac{2}{3} \frac{k}{(1 + k)^2} \frac{d_f^2 \bar{u}}{D_S} \quad (5.1)$$

where D_M and D_S represent diffusion coefficients in the mobile and stationary phases, respectively, f_1 and f_2 are pressure drop correction factors, k is the solute partition ratio, d is the column diameter, and $\bar{u}$ is the average linear velocity of the mobile phase.

Consideration of Eq. (5.1) makes it apparent that temperature effects will be interrelated and that the precise overall result of a temperature change is not easily defined. With most stationary phases, D_S varies directly with temperature (see Section 5.13), but there are some indications that this is not always true. A temperature change would also affect D_M, but the precise quantitative effect is not readily predictable: both gas viscosities and the kinetic energy of the solute molecules vary directly with temperature; the former would tend to decrease, and the latter to increase, the diffusivity of that solute in the mobile phase. Qualitatively, D_M usually increases with increasing temperature.

For a given solute, any shift in temperature would affect the partition ratio k and introduce additional factors, considered below. The changes in gas viscosity, gas density, and diffusivity of the solute molecules in both the stationary and mobile phases affect both the longitudinal diffusion (B) term and the resistance to mass transport (C_M and C_S) terms of the van Deemter (or Golay) equation, even if the solute is varied to maintain a constant k. The pressure drop correction factors in Eq. (5.1) were introduced to compensate at least partially for some of these complications.

5.6 Effect of Solute Partition Ratios on Optimum Gas Velocities

Figure 5.3 shows computerized plots representing the effects of the partition ratio of the test solute on the van Deemter curve under the conditions specified in the figure legend. Evaluation of the curves in terms of h_{min} confirms earlier considerations related to Fig. 1.9: the magnitude of h_{min} varies directly (and n

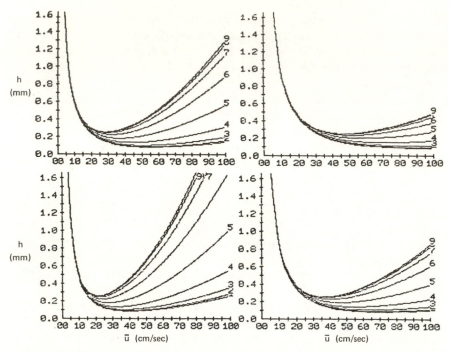

Fig. 5.3. Effect of the solute partition ratio on the van Deemter curve. Solute partition ratios increase by a factor of 3 for each determination (bottom to top): 0.01, 0.03, 0.09, 0.27, 0.81, 2.43, 7.29, 21.87, and 65.61. Upper figures, 30 m × 0.25 mm columns, d_f 0.25 μm. Lower figures, 60 m × 0.25 mm columns, d_f 0.25 μm. Left-hand figures, helium carrier (D_M 0.398). Right-hand figures, hydrogen carrier (D_M 0.561 cm²/sec). D_S is taken as 1.5×10^{-6} cm²/sec in all cases. Note that the differences imposed by a threefold increase in k are most pronounced in the region $k = 0.1–7$.

varies inversely) with the partition ratio of the test solute; i.e., low-k solutes exhibit small values for h_{min} (and high values for n), and high-k solutes exhibit larger values for h_{min} (and smaller values for n).

It is also apparent from Fig. 5.3 that $\bar{u}_{opt}$ varies inversely with the solute partition ratio: it occurs at higher velocities for low-k solutes and at lower velocities for higher-k solutes. This leads to the disturbing conclusion that even under isothermal conditions, the selection of a carrier gas velocity yielding optimum separation over the entire chromatogram will not be possible.

Isothermal Temperature Considerations

If there were one particularly demanding area in the chromatogram (e.g., one solute whose separation was especially important), the carrier gas velocity could be optimized for that particular value of k and the entire chromatogram run under those conditions; some degree of separation efficiency would be sacrificed at all other values of k.

Figure 5.4 shows plots of h versus $\bar{u}$ for nitrogen, helium, and hydrogen carrier gas, with k varied from 0.1 to 15. In all cases, what we might term the $d\bar{u}/dk$ ratio (the change in $\bar{u}_{opt}$ per incremental change in k) is greatest for low-k solutes. This relationship is explored in Fig. 5.5, where it can be seen that in the range $k = 0$–1.0, the magnitude of k has a profound effect on $\bar{u}_{opt}$; it is significant in the range $k = 1$–2, and for $k > 2$ the deviation is small.

A lower column temperature (which would decrease solute vapor pressures, hence increase solute distribution constants and partition ratios; see below) would, of course, be one way of achieving improved separation over a range of solute partition ratios that included very low-k solutes. In a pressure-controlled system, a lower column temperature would also result in an increased carrier gas velocity. However, because of restrictions discussed later, lower column temperatures are not always feasible.

The practical effects of the gas velocity decision can be demonstrated by considering the separation of a complex mixture whose solute partition ratios, under the analytical conditions that must be used, range from <0.5 to >10 (this range of solute partition ratios would not be unusual for a large number of natural products). Because $\bar{u}_{opt}$ varies with k, only a few of these solutes could be subjected to separation at or near their optimum velocities at any one value of $\bar{u}$. As an example, assume that the best possible separation efficiency is desired with a 30 m × 250 μm column, $d_f = 0.25$ μm, over the range $k = 0.1$–15 (Fig. 5.4). With nitrogen carrier gas, the lowest-k solute would exhibit an h_{min} of 0.1 mm at a $\bar{u}_{opt}$ of 36 cm/sec and the highest-k solute an h_{min} of 0.24 mm at a $\bar{u}_{opt}$ of 18 cm/sec. If the separation were performed at $\bar{u} = 18$ cm/sec (the optimum velocity for the high-k solute), the lowest-k solute would be subjected to (30 m/0.15 mm) = 200,000 theoretical plates (sacrificing 33% of the possible theoretical plates; see below) and the highest to (30 m/0.24 mm) = 125,000 theoretical plates. At 36 cm/sec (the optimum velocity for the low-k solute), the lowest-k solute would be subjected to (30 m/0.1 mm) = 300,000 theoretical plates and the highest-k solute to (30 m/0.4 mm) = 75,000 theoretical plates (sacrificing 40% of the possible theoretical plates). Because the curves are obviously asymmetric, selection of a carrier velocity based on the average of the two extremes would discriminate heavily against higher-k solutes.

Another approach would be to decide the maximum efficiency loss that can be tolerated at any one value of k in order to achieve a reasonable efficiency over a broader range of solute partition ratios. By allowing a 10% increase in the values of h at both extremes (i.e., $h = 0.11 + 0.01 = 0.12$ at $k = 0.15$ and $h = 0.24 + 0.02 = 0.26$ at $k = 12$), at least 90% of the available theoretical plates could be brought to bear at all points throughout the separation. As illustrated in Fig. 5.6, the velocity limits indicated by the intersection of these new values with the respective van Deemter curves establish a velocity window, whose lower limit is defined by the lower limit of the lower-k solute and whose upper limit is defined by the upper limit of the higher-k solute. In Fig. 5.7, this concept is applied to

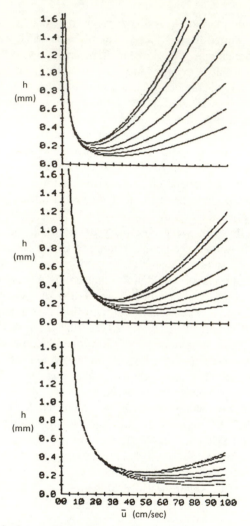

Fig. 5.4. Effect of solute partition ratio on van Deemter curves, operating with different carrier gases. All columns 30 m × 0.25 mm, d_f 0.25 μm. Top, nitrogen carrier (D_M 0.15 cm²/sec); middle, helium carrier (D_M 0.398 cm²/sec); bottom, hydrogen carrier (D_M 0.56 cm²/sec); partition ratios (from bottom curve to top curve): 0.1, 0.25, 0.5, 1.0, 3.0, 7.0, and 15. Other conditions are the same as in Figs. 5.3 and 5.4.

computer-constructed van Deemter curves for solutes $k = 0.1$ and $k = 10$, for nitrogen, helium, and hydrogen carrier. The velocity window is narrowest for nitrogen carrier, broader for helium carrier, and broadest for hydrogen carrier gas. Obviously, flatter van Deemter curves permit broader windows, less efficiency will be sacrificed by slight departures from the optimum velocity, and

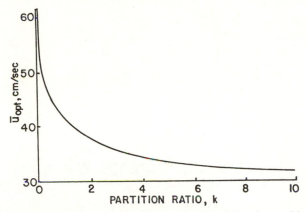

Fig. 5.5. Optimum average linear gas velocities as a function of the solute partition ratio. Calculated for a 30 m × 0.25 mm column, d_f 0.25 μm, helium carrier (D_M 0.398 cm²/sec), D_S at 1 × 10^{-6} cm²/sec.

broader ranges of solute partition ratios can be subjected to good separation at some one velocity.

Programmed Temperature Considerations

Some open tubular column systems operate at constant pressure and others operate at constant flow; gas viscosities vary directly with temperature. In pressure-regulated systems, this results in decreased carrier gas flows as the program proceeds to higher temperatures, where higher-k solutes, which exhibit lower

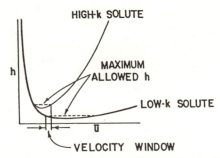

Fig. 5.6. The "velocity window" concept. The analyst must first decide the maximum increase in h to be tolerated at the two limits of interest in the chromatogram (lowest-k solute and highest-k solute) between which a high degree of separation is desired. The window is defined by the values of u corresponding to the upper limit of the high-k (low optimum velocity) solute and the lower limit of the low-k (high optimum velocity) solute. If the upper limit of the high-k solute is lower than the lower limit of the low-k solute, no window exists for the conditions specified; the analysis must be limited to a narrower range of solutes, or a greater increase in h must be permitted.

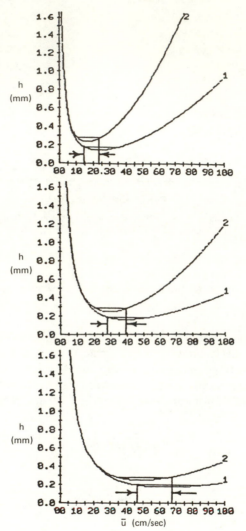

Fig. 5.7. The velocity window concept applied to computer-constructed van Deemter curves for solute partition ratios $k = 0.5$ and $k = 10$ on a 30 m × 0.25 mm column, d_f 0.25 μm; top, nitrogen carrier; middle, helium carrier; bottom, hydrogen carrier. Other parameters as in Figs. 5.3 and 5.4; a 10% increase in h_{min} has been allowed. Note that as the van Deemter curves become flatter, windows occur at higher velocities and become broader.

values of $\bar{u}_{opt}$, elute. Although the magnitude of the gas flow shift is open to question, it is at least in a compensatory direction.

Our considerations to this point indicate that a pragmatic approach should be possible in deciding on a flow velocity for a programmed temperature separation. Assuming that there is no single critical region, the velocity is probably best set at the $\bar{u}_{opt}$ for higher-k solutes at or up to 30°C below the temperature at which they elute. At lower temperatures, the velocity will be much higher, but the lower-k solutes exhibit higher optimum velocities, and their van Deemter curves are also flatter; particularly where it errs in being on the high side, low-k solutes suffer less for a given departure from their $\bar{u}_{opt}$ than do high-k solutes.

5.7   Effect of Solute Partition Ratios on Separation Potentials

The number of theoretical plates required to achieve a given resolution on two solutes is governed in part by the partition ratio k of the second solute [Eq. (1.22)]. To what degree the $[(k + 1)/k]^2$ multiplier is affected by the magnitude of k is explored in Table 5.1: increasing k by an order of magnitude reduces n_{req} by 99.4% at $k = 0.01$, by 97.7% at $k = 0.1$, by 84.5% at $k = 0.5$, 72% at $k = 1.0$, and about 25% at $k = 5.0$. The examples indicate that the separation of very low-k solutes will benefit greatly, that of intermediate-k solutes moderately, and that of higher-k solutes little or not at all from a given increase in solute partition ratios.

One route to increased partition ratios is through stationary phase selectivity; although the primary benefit of a more retentive stationary phase (i.e., one with increased solute–stationary phase interactions) is usually evidenced by larger

TABLE 5.1

Values of $[(k+1)/k]^2$ and n_{req} Corresponding to Different k Values

k	$[(k+1)/k]^2$	$n_{req}{}^a$	$n_{req}{}^b$
0.01	10,201	162,000,000	44,000,000
0.05	441	7,000,000	1,900,000
0.10	121	1,900,000	527,000
0.15	58.8	930,000	256,000
0.5	9	143,000	39,000
1.0	4	63,500	17,500
3.0	1.8	28,600	7,800
5.0	1.4	22,200	6,100
10.0	1.1	17,500	4,800
50.0	1.05	16,700	4,600

Plate numbers have been rounded and assume $Rs = 1.5$ and ${}^a\alpha = 1.05$, ${}^b\alpha = 1.10$.

relative retentions (α), solute partition ratios are also higher in the more retentive stationary phase, leading to higher values of k. There are two other routes to increased solute partition ratios: lower column temperatures (or lower program rates) and columns of lower phase ratio.

Solute vapor pressures are much depressed at lower column temperatures. As

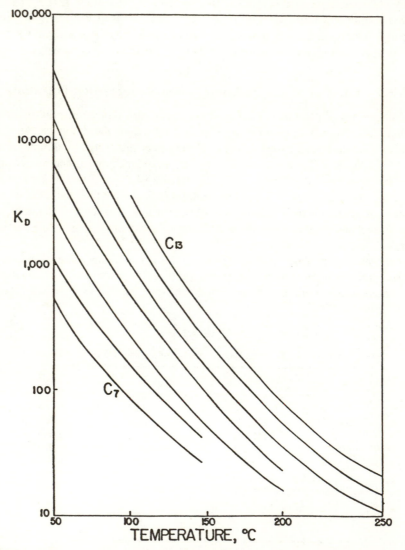

Fig. 5.8. Effect of temperature on the distribution constants of selected n-paraffin hydrocarbons in a dimethylpolysiloxane stationary phase. (Replotted from data of Butler and Hawkes [10]).

a result, c_S values increase and c_M values decrease; the increase in K_D forces a proportionate increase in k [Eq. (1.14)]. Figure 5.8 illustrates the effect of temperature on the distribution constants of some normal alkanes in a polysiloxane stationary phase [10], and Table 5.2 lists values for several solutes in several stationary phases and at different temperatures [11,12].

If K_D and k are large at the time that the solute reaches the end of the column, the solute elutes to the detector at lower concentrations over a longer period of time. As discussed in Chapter 1, this causes broadened peaks, which affects sensitivity adversely. Under conditions of temperature programming, the K_D of each solute changes continuously during its passage through the column. Program parameters (initial and final temperatures, initial and final holds, number of ramps, and program rates) and carrier flow conditions should be in balance with column parameters (length, phase ratio, and stationary phase) to mandate larger values of K_D—and smaller values for the $[(k + 1)/k]^2$ multipliers—during most of the analysis, and smaller values of K_D (hence improved sensitivities) as each solute reaches the end of the column and is delivered to the detector. Because it

TABLE 5.2

Selected Distribution Constants[a]

	Stationary phase								
	SE-30		OV-210			OV-225		PEG 20M	
Solute	51°C	100°C	48°C	97°C	149°C	97.5°C	148°C	98°C	148.5°C
Hexane	96								
Heptane	229		66						
Octane		87	139						
Nonane			302	45					
Decane			671	80		98			
Undecane				143		183	31		
Dodecane				255	46	348	52		
Tridecane				463	71		82		
Tetradecane					108		134		
Methanol								32	9
Ethanol								35	10
Propanol								64	15
Butanol								113	24
Pentanol								231	38
Hexanol								400	61
Octanol								1192	139
Benzene	161	37	117			42			
Chloroform	102		64			38			

[a]Data selected from Hawkes [11,12].

permits manipulation of the solute K_D in this manner, temperature is the most elegantly simple route to controlling the solute partition ratio. Normally, only those cases where either the temperature constraints of the stationary phase (either minimum or maximum temperature limits) or instrumental restrictions (e.g., lack of subambient capability) preclude operation of the column at the required temperatures justify using one of the other routes to the manipulation of k (different stationary phase, different phase ratio).

5.8   Effect of Column Length

Figure 5.9 shows plots of different-length Megabore columns with hydrogen, helium, and nitrogen carrier gas for various solute partition ratios. Differences in the behavior of 15- and 30-m lengths are relatively slight with these large-diameter, low pressure drop columns. Figure 5.10 examines the effect of column length on 0.32 mm i.d. columns with both helium and hydrogen carrier gas; "Minibore" (0.15 mm i.d.) and "Microbore" (0.05 and 0.10 mm i.d.) columns are examined in Figs. 5.11–5.13. Note that for a given partition ratio, $\bar{u}_{opt}$ varies indirectly with column length. If each column were used at the optimum average linear carrier gas velocity for some one solute with all other conditions remaining the same, the longer the column, the more column the solutes must traverse, and the lower their velocity per unit length of column; i.e., analysis times become disproportionately longer as the column length increases. The major reason for using longer columns is to subject the solutes to a larger number of theoretical plates, but component resolution increases only with the square root of column length and then only at $\bar{u}_{opt}$. The other disadvantage of longer columns is related

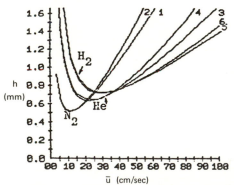

$\bar{u}$ (cm/sec)

Fig. 5.9. Effect of column length on large-diameter open tubular columns as examined by computer-constructed van Deemter curves. All columns 0.53 mm i.d., d_f 1.0 μm, $k = 3$. (1) 15-m column, nitrogen carrier; (2) 30-m column, nitrogen carrier; (3) 15-m column, helium carrier; (4) 30-m column, helium carrier; (5) 15-m column, hydrogen carrier; (6) 30-m column, hydrogen carrier. Other parameters as in Figs. 5.3 and 5.4. The effects of increased column length are less pronounced with these large-diameter (low pressure drop) columns.

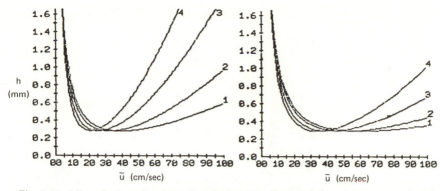

Fig. 5.10. Effect of column length on ''standard capillary'' open tubular columns as examined by computer-constructed van Deemter curves. Solute partition ratio 3.0; columns 0.32 mm i.d. and d_f 0.25 μm. Left, helium carrier; right, hydrogen carrier. Column length: (1) 15 m, (2) 30 m, (3) 60 m, and (4) 100 m. Other parameters as in Figs. 5.3 and 5.4.

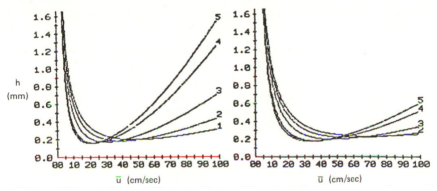

Fig. 5.11. Effect of column length on columns of the Minibore series. Solute partition ratio 3.0; columns 0.15 mm i.d., d_f 0.4 μm. Left, helium carrier; right, hydrogen carrier. Column length: (1) 5 m, (2) 10 m, (3) 20 m, (4) 40 m, and (5) 50 m. Other parameters as in Figs. 5.3 and 5.4.

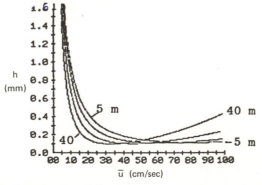

Fig. 5.12. Effect of column length on Microbore columns. Solute partition ratio 4.0; columns 0.10 mm i.d., d_f 0.2 μm. All curves with hydrogen carrier. Column length: (1) 5 m, (2) 10 m, (3) 20 m, and (4) 40 m. Other parameters as specified in Figs. 5.3 and 5.4.

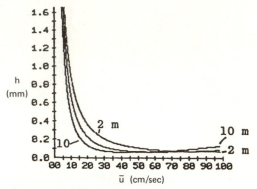

Fig. 5.13. Effect of column length on Microbore columns. All conditions as in Fig. 5.12, except column diameters are 0.05 mm i.d. and column lengths are (1) 2 m, (2) 5 m, and (3) 10 m. The high pressure drops of these small-diameter columns require hydrogen inlet pressures exceeding 80 psi for the operation of a 10-m column at u_{opt}, and helium becomes an impractical carrier.

to the pressure drop through the system. In general, a longer column requires a larger pressure drop (column diameter and carrier gas held constant), resulting in a steeper van Deemter curve, with the disadvantages discussed in Section 5.6. Ideally, the column length should be just sufficient to provide the number of theoretical plates required for the separation, and no longer.

5.9   Effect of Column Diameter

Figure 5.14 explores the effect of varying the column diameter for a fixed length of column, while holding d_f constant. The resultant change in the column phase ratio would shift the solute partition ratios as indicated in the figure legend. In Fig. 5.15 both diameter and d_f are varied (constant length) to hold the phase ratio (and solute partition ratio) constant. Smaller-diameter columns, of course, generate higher pressure drops per unit length, resulting in steeper van Deemter curves. However, from Eq. (5.1), it is obvious that the theoretical plate number varies indirectly with the column diameter:

$$h_{\text{theoretical minimum}} = r[(11k^2 + 6\,k + 1)/3(1 + k)^2)]^{1/2}$$

This makes it evident that the comparisons shown in Figs. 5.14 and 5.15 are unrealistic; at a constant length, smaller-diameter columns generate more theoretical plates. In Fig. 5.16, both the column length and the stationary phase film thickness (d_f) have been varied commensurate with column diameter to maintain a constant column phase ratio and a constant number of theoretical plates.

There are other factors that must be considered in conjunction with column diameter; although microbore columns are capable of generating more plates per meter, they impose additional demands on the operator and on the equipment.

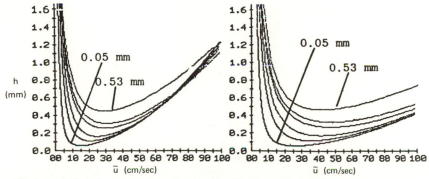

Fig. 5.14. Effect of varying column diameter, holding d_f constant at 0.4 μm; the resulting change in the phase ratio of the column would, at constant temperature, force shifts in solute partition ratios as indicated below. All columns 30 m, other parameters as indicated in Figs. 5.3 and 5.4. Left, helium carrier; right, hydrogen carrier. Column diameters and solute partition ratios (bottom curves to top curves): 0.05 mm and k = 25; 0.10 mm and k = 12.5; 0.15 mm and k = 8.3; 0.25 mm and k = 5.0; 0.32 mm and k = 3.9; 0.53 mm and k = 2.4.

Larger-diameter columns (e.g., 530 μm) appeal to many users because their high flow volumes yield a "forgiving system" that can be directly interfaced to most packed column injectors and detectors (they also have increased sample loading capacities). By the same reasoning, interfacing true microbore columns (e.g., 50 μm i.d.) to conventional inlets and detectors may be extremely difficult; such columns are easily overloaded, and their very low flow volumes (~50 μl/min) require split ratios as high as 1 : 2000 to flush the inlet clean and limit the initial

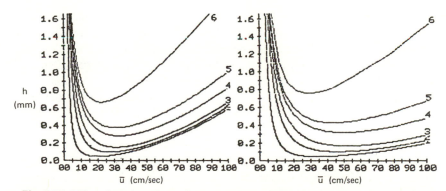

Fig. 5.15. Effect of varying column diameter while varying d_f to hold the column phase ratio (and solute partition ratio) constant. Left, helium carrier; right, hydrogen carrier. Column length constant at 20 m; the number of theoretical plates generated under these conditions for a solute k = 3 would be 255,000 at 0.05 mm (bottom curves), 128,000 at 0.10 mm, 85,000 at 0.15 mm, 51,000 at 0.25 mm, 40,000 at 0.32 mm, and 24,000 at 0.53 mm (top curves).

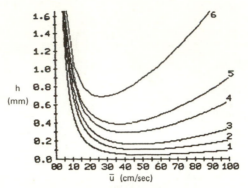

Fig. 5.16. Computer-constructed van Deemter curves calculated for columns of various diameters but constant phase ratio, with column length adjusted to generate 100,000 theoretical plates on a solute $k = 4.0$. All systems had a hydrogen carrier; other parameters were as indicated in Figs. 5.3 and 5.4. Curves: (1) 8.2 m $\times$ 0.05 mm, d_f 0.1 μm; (2) 16.4 m $\times$ 0.10 mm, d_f 0.2 μm; (3) 24.5 m $\times$ 0.15 mm, d_f 0.3 μm; (4) 41 m $\times$ 0.25 mm, d_f 0.5 μm; (5) 52.4 m $\times$ 0.32 mm, d_f 0.6 μm; (6) 87 m $\times$ 0.53 mm, d_f 1.1 μm.

sample band length. Although it is not a simple matter to produce acceptable results with microbore columns, some interesting results have been reported (e.g., [13,14]). Figure 5.17 shows how analysis times can be shortened by using shorter microbore columns instead of standard capillary systems. These relationships are discussed further in later chapters.

5.10   Effect of Stationary Phase Film Thickness

Grob and Grob were among the first to explore the practical significance of varying d_f, particularly with thicker-film columns [15], but the thick-film columns of those early times had extremely short lifetimes. With the advent of cross-linked and bonded stationary phases (see Chapter 4), stable thick-film columns became possible. Open tubular columns with 1-μm films were commercially available in 1980, and additional efforts on ultrathick-film columns [16–19] soon led to their commercial production. Today, columns can be obtained with stationary phase film thicknesses of 0.1–0.15, 0.2–0.4, 0.5, 1.0, 1.5, 3, 5, and 8 μm.

At constant temperature, the solute partition ratio k varies indirectly with the column phase ratio β [Eqs. (1.17) and (1.81)]. Column phase ratios can be increased by using columns of smaller diameter or of increased stationary phase film thickness. If K_D is constant (same solute, stationary phase, temperature), the effects of such changes on solute partition ratios can be predicted by rewriting Eq. (1.17) in the form

$$k_{(2)} = [r_{(1)}k_{(1)}d_{f(2)}]/[r_{(2)}d_{f(1)}] \tag{5.2}$$

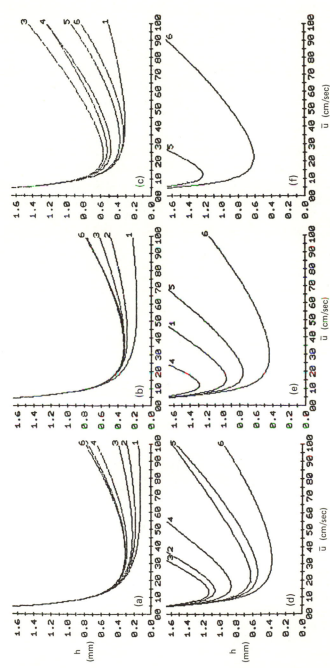

Fig. 5.17. Effect of varying the thickness of the stationary phase film on "standard capillary" columns. Columns 15 m × 0.32 mm, helium carrier; D_M 0.398 and D_S 1.5 × 10^{-6} cm^2/sec. Solute partition ratios (each graph): (1) 0.1, (2) 0.5, (3) 1.0, (4) 5.0, (5) 20, (6) 100. Values of d_f (μm): (a) 0.2, (b) 0.5, (c) 1.0, (d) 2.0, (e) 3.0, (f) 5.0. Note that with thin films there exists a regular progression in the magnitude of the increase in h_{min} with increasing k; at d_f values exceeding ∼0.4, curves are displaced upward and become steeper, and low-k curves are more affected than high-k curves. With extremely thick films, high-k solutes finally exhibit lower values for h_{min} than do low-k solutes.

The phase ratios of many commercially available columns are shown in Table 5.3.

The reduction in n_{req} (Table 5.1) occasioned by the increase in d_f must be balanced against the fact that as d_f is increased beyond about 0.4 μm, h_{min} increases [20] and $(\bar{u}_{opt}$ decreases, Section 5.8). In other words, although the thicker-film column increases solute partition ratios and so achieves a given degree of resolution with fewer theoretical plates, fewer theoretical plates are delivered by the thicker-film column. Whether the net result is positive or negative for the separation of a given solute pair depends on (1) the increase in the partition ratio of the second solute and (2) the magnitude of the increase in h_{min} (i.e., the loss in column efficiency). The benefits (in terms of the reduction in n_{req}) are most evident for very low-k solutes. Increasing d_f from 0.25 to 2.5 μm would increase solute partition ratios by an order of magnitude (all other factors constant). As shown in Table 5.1, this would have a profound effect on n_{req} for low-k solutes (e.g., $k = 0.01$); it would yield some benefits for solutes in the

TABLE 5.3

Phase Ratios of Selected Columns

d (mm)	r (μm)	d_f (μm)	B^a
0.05	25	0.2	63
0.10	50	0.4	63
0.15	75	0.5	75
0.20	100	0.4	125
0.25	125	0.1	625
0.25	125	0.15	416
0.25	125	0.25	250
0.25	125	0.5	125
0.25	125	1.0	63
0.32	160	0.1	800
0.32	160	0.15	533
0.32	160	0.25	320
0.32	160	0.5	160
0.32	160	1.0	80
0.32	160	3.0[b]	27
0.32	160	5.0[b]	15
0.53	265	1.0	133
0.53	265	1.5	88
0.53	265	3.0[b]	44
0.53	265	5.0[b]	27

[a]Rounded to the nearest whole number.
[b]These very thick film columns confer other disadvantages and should be restricted to high-diffusivity phases; see Sections 5.10 and 5.11.

range $k = 0.5-1$; for solutes of $k = 5$ and higher, the advantages are negligible or nonexistent and are usually outweighed by the disadvantages of the thicker film. The picture is complicated by the fact that the rate at which h_{min} increases with k is less pronounced for high-k solutes than for low-k solutes in these very thick film columns. This relationship is discussed later.

While the overall effect of the stationary phase film thickness will be complicated by the fact that any change in d_f will also affect k, Eq. (5.1) makes it evident that thicker films of stationary phase will have a deleterious effect on column efficiency. The effect was experimentally recognized even before the advent of the van Deemter treatment [6]. The columns that were used in the very early days of gas chromatography were frequently packed with crudely crushed firebrick, whose inadequate deactivation was masked by thicker coatings of stationary phase; it was not unusual to use packings that contained 20–25% by weight stationary phase. With better deactivation procedures, lighter loadings became possible, and these delivered improved column efficiencies as measured by the theoretical plate number. Probably because of these considerations, a significant fraction of gas chromatographers seem convinced that open tubular columns with thinner films must give greater separation efficiencies. The influence of stationary phase diffusivity is discussed below, but with the higher-diffusivity phases, d_f exercises little effect on the magnitude of h_{min} until the thickness of the stationary phase film exceeds about 0.4 μm. Indeed, because of the relationships explored below [Eq. (5.2) and Table 5.1], the decrease in solute partition ratios occasioned by thinner films may increase n_{req} to such a degree that resolution is affected adversely.

The major reason for varying the stationary phase film thickness is to change the column phase ratio, which has effects on column capacity, solute partition ratios, solute separation, analysis times, and sensitivities (Section 5.7).

If the column radius is held constant, the solute partition ratio varies directly with the thickness of the stationary phase film. Thin-film columns find their greatest utility in the analysis of high-boiling solutes, which would exhibit very long retentions and lowered sensitivities on normal (or thick-film) columns. Even with well-deactivated fused silica columns, excessive activity is displayed with stationary phases such as the phenylcyanopropylmethyl silicones at d_f values less than 0.15 μm, implying that the surface "shows through." Dimethyl silicones apparently give better coverage and can produce satisfactory columns at film thicknesses of 0.1 μm. Thin-film columns can generate inflated values when used for the determination of retention indices; the adsorptive effects that are contributed by even well-deactivated surfaces (and which have little effect on paraffin hydrocarbons but retard polar solutes) tend to "show through" thin films.

Thick-film columns have two major advantages and several disadvantages. The advantages are that (1) larger or more concentrated samples can be handled before overloading occurs [21], and (2) solute partition ratios become larger, which can be beneficial in the separation of some solutes. The latter advantage can be attributed to the higher values of the solute partition ratio k and to the effect that this has on the magnitude of the $[(k + 1)/k]^2$ multiplier of Eq. (1.23).

The ultrathick-film column is extremely valuable for the separation of very low-k solutes (see Chapter 9), and its great capacity has excited some interest in those using techniques such as gas chromatography–Fourier transform infrared (GC–FTIR). Because of the relationships cited above, it should not be regarded as a general-purpose column.

There are many reports that bleed rates are higher with thick-film columns. As detailed in Chapter 10, column bleed rates can be attributed to several sources. With thoroughly end-capped and ultraclean polysiloxane-type stationary phases and oxygen-free carrier, bleed rates can be significantly lower, but they still correlated reasonably well with the mass of stationary phase.

Figure 5.18 shows the effect of varying d_f in 15 m $\times$ 0.32 mm columns. Note that as d_f increases, all curves are displaced upward and become steeper; for a given increase in d_f, low-k solutes are more affected than high-k solutes. The solutes $k = 0.1$ and $k = 20$ exhibit values of h_{min} of 0.13 and 0.31, respectively, at d_f 0.2 μm; at d_f 3.0 μm the values of h_{min} have increased to 0.96 and 0.76 for $k = 0.1$ and 20, respectively. At 5 μm, the only solutes still on scale are $k = 20$ and $k = 100$. Figure 5.19 explores the same relationships on 15 m $\times$ 0.53 mm i.d. columns, where "crossover" occurs at the same value of d_f.

5.11   Effect of Stationary Phase Diffusivity

The importance of stationary phase diffusivity is closely correlated with stationary phase film thickness; both influence the mass transport term, C_S, of the van Deemter equation. It would seem reasonable to expect the viscosity of a stationary phase to yield some indication of diffusivity in that phase, but this has proved untrue in at least one example. Data from studies supervised by Hawkes [10–12] indicate that diffusivities are highest in the polysiloxane gums (SE-30) and are in most cases slightly lower in the polysiloxane fluids (SF-96). This apparent discrepancy has been interpreted as an indication of a more porous, open structure of the gum. Substitution of other groups for some of the methyl decreases diffusivities; in general, Hawkes' data indicate that diffusivities in some of the common stationary phases vary directly with temperature and at any one temperature approximate the order CH_3 > phenyl > trifluoropropyl > polyethylene glycol > cyanopropyl.

In cross-linked stationary phase (see Chapter 4), diffusivities (and, of course, viscosities) can be affected by the extent of cross-linking.

The combined significance of d_f and D_S is indicated in Fig. 5.19. Scanning across the top figures indicates that stationary phase diffusivity contributes little

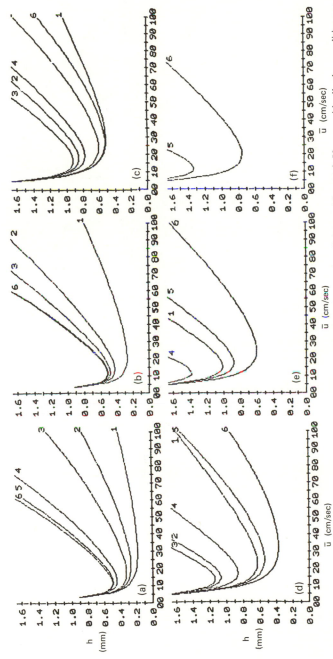

Fig. 5.18. Effect of stationary phase film thickness on columns of the Megabore series. Columns 15 m × 0.53 mm, with all other conditions as specified in Fig. 5.17. Note that the various "crossover points" occur as functions of d_f and not of the column phase ratio.

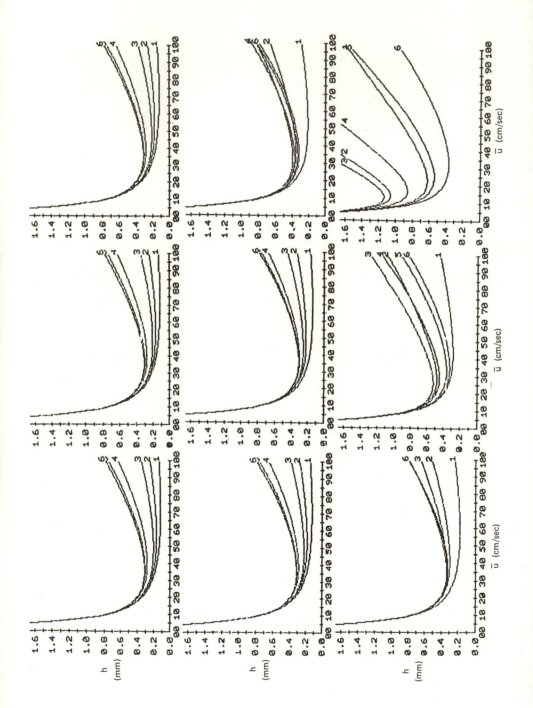

in thin-film ($d_f < 0.4$ μm) columns. Similarly, the left-hand figures indicate that the effect of d_f is minimized by high-diffusivity phases. With lower D_S and/or d_f, all curves become steeper and are displaced upward (h_{min} increases). Because of these factors, most manufacturers prefer to limit columns with very thick films ($d_f > 2$ μm) to the high-diffusivity dimethyl polysiloxane stationary phases. Specially synthesized lower-viscosity polymers, whose cross-linking is carefully regulated, have been found to yield better efficiencies with stationary phase films of 3 and 5 μm [22].

5.12   Changes in Solute Elution Order

A surprising number of inquiries from practicing chromatographers are concerned with the fact that the solute elution order of a given sample changed with the substitution of another column coated with the same stationary phase. Solute–stationary phase interactions that influence solute volatility are considered in Chapter 4; the nature of the solute (and its functional groups) would be expected to affect the degree to which a temperature increase altered those interactions and hence the degree to which the volatility of the solute was affected by the temperature increase.

In a column coated with a stationary phase capable of a high degree of hydrogen bonding, a short-chain alcohol will be retained and elute after a hydrocarbon of much greater chain length when the mixture is chromatographed on that column at some given temperature T_1. This, of course, is because the alcohol is more susceptible to hydrogen bonding. If the same mixture is chromatographed on the same column at a significantly higher temperature T_2, the degree of hydrogen bonding is reduced and the alcohol may now elute before the hydrocarbon. Hence, the elution order at T_1 is alcohol–hydrocarbon, and at T_2 it becomes hydrocarbon–alcohol.

In a programmed temperature separation that begins at a temperature at or below T_1 and proceeds at a steady rate to a final temperature at or above T_2, the elution order will be determined by the length of time the solutes are subjected to the chromatographic process (i.e., the temperature profile that they experience). Anything that reduced the analysis time (higher gas velocities, shorter columns, larger-diameter columns at constant d_f, thinner stationary phase films at constant column diameter) would discharge the solutes at lower temperatures and tend to

---

Fig. 5.19. Effect of stationary phase diffusivity (D_S) for various solute partition ratios. Columns 15 m × 0.32 mm, d_f (top graphs) 0.2 μm, (center graphs) 1.0 μm, (bottom graphs) 3.0 μm; helium carrier. D_S calculated at: left, 5×10^{-5} cm²/sec; center, 1.6×10^{-6} cm²/sec; right, 3.4×10^{-6} cm²/sec. The higher value approaches that derived for naphthalene in SE-30 at 243°C, and the lowest that for dodecane in OV-225 at 150°C [10]. Solute partition ratios: (1) 0.1, (2) 0.5, (3) 1.0, (4) 5.0, (5) 20, (6) 100.

preserve the elution order alcohol–hydrocarbon. Anything that increased the analysis time (lower gas velocities, longer columns, smaller-diameter columns at constant d_f, or thicker stationary phase films at constant column diameter) would retain the solutes until the program had advanced further; the solutes would be exposed to higher temperatures and would tend to elute in the order hydrocarbon–alcohol. Obviously, a different program rate or different initial or final temperatures could also affect elution order.

Analysis of a mixture containing the pesticides parathion and aldrin can serve as an example. In a series of isothermal analyses conducted on an apolar (DB-5) column, the elution order on this column under the operational conditions employed was aldrin–parathion at 185°C; the two solutes coeluted as a single peak at 190°C; and at 195°C the elution order was parathion–aldrin. In a programmed run, their elution order would depend on the factors discussed above.

Programmed elution orders of solute mixtures not restricted to members of a homologous series would be expected to vary if changes were made in the length, diameter, or stationary phase film thickness of the column, in the carrier gas velocity (as happens when optimizing with hydrogen versus helium), in the pressure drop through the column, or in any of the program parameters.

References

1. J. C. Giddings, *Anal. Chem.* **36,** 741 (1964).
2. J. C. Sternberg, *Anal. Chem.* **36,** 921 (1964).
3. G. Takeoka, H. M. Richard, M. F. Mehran, and W. Jennings, *J. High Res. Chromatogr.* **6,** 145 (1983).
4. R. P. W. Scott and G. S. F. Hazeldean, *in* "Gas Chromatography 1960" (R. P. W. Scott, ed.), p. 144. Butterworth, London, 1960.
5. D. F. Ingraham, C. F. Shoemaker, and W. Jennings, *J. High Res. Chromatogr.* **5,** 227 (1982).
6. J. J. van Deemter, F. J. Zuiderweg, and A. Klinkenberg, *Chem. Eng. Sci.* **5,** 271 (1956).
7. J. C. Giddings, S. L. Seager, L. R. Stucki, and G. H. Stewart, *Anal. Chem.* **32,** 867 (1960).
8. J. C. Giddings, *Anal. Chem.* **35,** 314 (1962).
9. C. A. Cramers, F. A. Wijnheymer, and J. A. Rijks, *J. High Res. Chromatogr.* **2,** 329 (1979).
10. L. Butler and S. Hawkes, *J. Chromatogr. Sci.* **10,** 518 (1972).
11. J. M. Kong and S. J. Hawkes, *J. Chromatogr. Sci.* **14,** 279 (1976).
12. W. Milliken and S. Hawkes, *J. Chromatogr. Sci.* **15,** 148 (1977).
13. C. P. M. Schutjes, E. A. Vermeer, G. J. Scherpenzeel, R. W. Bally, and C. A. Cramers, *J. Chromatogr.* **289,** 157 (1984).
14. F. I. Onuska, *J. Chromatogr.* **289,** 207 (1984).
15. K. Grob, Jr. and K. Grob, *Chromatographia* **10,** 250 (1977).
16. K. Grob and G. Grob, *J. High Res. Chromatogr.* **6,** 133 (1983).
17. P. Sandra, I. Temmerman, and M. Verstappe, *J. High Res. Chromatogr.* **6,** 501 (1983).
18. L. S. Ettre, G. L. McClure, and J. D. Walters, *Chromatographia* **17,** 560 (1983).
19. P. Benedek, L. Jozsa, and L. S. Ettre, *Chromatographia* **18,** 367 (1984).
20. W. Jennings, *Am. Lab.* **16,** 14 (1984).
21. K. Yabumoto, D. F. Ingraham, and W. Jennings, *J. High Res. Chromatogr.* **3,** 248 (1980).
22. R. Jenkins and R. Lautamo (J & W Scientific, Inc.), personal communication (1985).

CHAPTER 6

COLUMN SELECTION, INSTALLATION, AND USE

6.1   General Considerations

Column selection is all too often historically based: someone once performed a certain separation on a given stationary phase, ergo that stationary phase should be used for that separation. Although the knowledge of separations that have been achieved in the past can provide some useful information, those separations should usually be considered as points of reference, whose results can probably be improved on in light of more recent developments. Columns today are more efficient and better deactivated; a wider range of stationary phase film thicknesses and column diameters extends our capabilities to manipulate solute partition ratios; newly developed stationary phases permit greater selectivity through adjustments of the relative retentions of key solutes; new techniques of injection may also be appealing, and some of these are practical only with bonded-phase columns.

Alternatively, the initial point of reference can take the form of a preliminary separation on a short apolar column, swept to its upper temperature limit at a high program rate under conditions of high carrier gas velocities. Routes to improving the analysis can be deduced from those preliminary results. The initial starting temperature can be lowered to improve the separation of the early-eluting components, and final temperatures can be lowered as the elution temperatures of the final compounds are established. Lower program rates will improve overall separation, and gas velocities can be more nearly optimized at u_{opt} or OPGV for the more critical region. In short, some preliminary information in the form of an initial chromatographic separation is required as a point of departure from which one can postulate the effect(s) that increasing column length, decreasing column diameter, changing d_f, or changing the stationary phase will have on the various areas of that original chromatogram.

The rationale in proceeding from either of these points of reference should be based on theoretical considerations that can be gleaned from Chapters 1, 4, and 5. As used in this section, those considerations have been complemented by long exposure to a vast number of chromatographic problems, questions, and observations from a multitude of users over a wide range of endeavor. It should also be pointed out that the installation of the column, including the type of seals and hanger design, can affect both the performance and the longevity of the column.

6.2   Selecting the Stationary Phase

Solute–Stationary Phase Incompatibility

Some columns have been specifically prepared for the separation of certain classes of solutes and may be unsuitable for some other classes of solutes. This mismatch is sometimes attributable to incompatibility of the solutes and the stationary phase and at other times to the use of surface pretreatments or deactivation procedures that render that column unsatisfactory for certain solutes or types of analysis. The "Carbowax-amine-deactivated" (CAM) columns can serve as an example. A strongly alkaline column is required for amine analysis; the tubing, which may or may not be deactivated [polyethylene glycol (PEG) is often allowed to serve as its own deactivating agent], is usually flushed with aqueous KOH, dried, and coated with Carbowax 20M. It is obvious that acidic solutes will be unable to negotiate these strongly alkaline columns. Columns of this type should always be flame sealed when not in use: KOH is hygroscopic, and Carbowax is water-soluble; even under the best of conditions, the lifetime of a standard CAM column will not be extensive. While some cross-linked and surface-bonded forms of PEG do not dissolve in water, the strongly alkaline substrate inhibits the bonding reactions; bonded CAM columns are not feasible at our present state of knowledge.

An analogous incompatibility can be demonstrated between columns coated with the "free fatty acid phase" (FFAP) or OV-351 stationary phases and slightly alkaline solutes. These columns are usually prepared by refluxing Carbowax 20M with terephthalic acid under conditions where each (hopefully) of the terminal OH groups of the polyethylene glycol condenses with one of the carboxyl moieties of the difunctional acid, leaving the other carboxyl free. The resulting polymer is acidic, and alkaline solutes are unable to negotiate the column. Again, the acidic reaction inhibits bonding. In order to utilize nonextractable columns for acidic solutes, however, some users pretreat columns coated with one of the bonded polyethylene glycols to temporarily enhance their performance with slightly acidic solutes. Typically, the column is rinsed with a few milliliters of dichloromethane containing 10 ppm FFAP or OV-351 (dilute phosphoric acid has also been employed) and dried under carrier flow to impart a slightly acidic reaction to the column. The eventual departure of the FFAP residues dissolved in the bonded PEG phase is accelerated by exposure to higher temperatures or by prolonged use, and the procedure must be repeated.

In some cases, a typical "overloaded" peak really denotes incompatibility between the solute and the stationary phase. Overloaded peaks, which are discussed in Chapter 10, are characterized by a sloping peak front and a more abrupt drop at the rear of the peak. Such peaks are generated when the equilibrium K_D cannot be established because the solute vapor pressure is too low to permit vaporization of all of that solute not dissolved by the stationary phase [1]. The problem can be corrected by increasing the solute vapor pressure (higher temperature, which also lowers K_D) or by increasing the solute-dissolving capacity of the stationary phase (thicker films of stationary phase lower the amount that must be accommodated by the gas phase at the equilibrium K_D). When equal amounts of an extended homologous series are subjected to an isothermal separation, the later peaks are often overloaded; vapor pressures of the higher-boiling (later) solutes limit the amount of those solutes that can be in the vapor phase. Solutes that are essentially insoluble in a given stationary phase are normally forced by that incompatibility to generate overloaded peaks on that phase; even small amounts of free acids typically yield overload peaks on methyl silicone phases (Fig. 3.5).

Nonbonded Stationary Phases

If precise retention indices [2] are of major concern, the stationary phase has been partially defined by that requirement; most precision indices utilize one of the dimethyl polysiloxane stationary phases (OV-1, SE-30) as one standard and a polyethylene glycol such as Carbowax 20M for the other [3]; indices determined on the former are, of course, far more precise. The PEG phases are based on average molecular weights (Carbowax 20M, MW_{av} 20,000; range not specified), and batch-to-batch variations in the concentration and availability of functional groups are the rule rather than the exception. In addition, PEG reacts readily with oxygen (and with some solutes), and the retention characteristics of these columns often change with use and with their operational history.

Because of these considerations, retention index windows can be quite narrow ($\pm 0.1\%$) on SE-30 and must be appreciably wider (e.g., ± 0.5 to $1+\%$) on a PEG phase. Bonding reactions require the incorporation of vinyl into the stationary phase oligomers prior to polymerization; cross-linking consumes some of the vinyl moieties, and others participate in bonding with pretreated silanol groups on the silica surface. Slight batch-to-batch variations may occur in the extent of these two reactions, and there is also a slight possibility of residual vinyl functionality. These variations would have some effect on the retention characteristics. Retention indices on bonded-phase columns are consequently slightly less precise and must vary to some degree from those obtained on the nonbonded counterpart. But even where retention characteristics are of primary importance, the advantages conferred by a bonded-phase column (see below) often outweigh the fact that retention windows must be broadened to employ the bonded phase.

In most practical situations and certainly whenever splitless or on-column injections will be employed, there are definite advantages to bonded phase columns. Not only will the bonded phase column resist the localized phase stripping that inevitably accompanies splitless and on-column injections, but nonvolatile and high-boiling residues can be cleansed from a bonded phase column by washing with an appropriate solvent. This "cleanability" helps in prolonging column lifetime and is an important preliminary step to column rejuvenation treatments that involve end-capping reactions (Chapter 10).

It is also good general practice to use the least polar stationary phase that is capable of producing the required separation; apolar columns usually exhibit higher efficiencies (lower h^*), have lower minimum and higher maximum operating temperature limits, have longer lifetimes, and are more amenable to rejuvenation treatments. Columns coated with bonded forms of dimethyl polysiloxane (e.g., DB-1; Ultra-1) or SE-54 (e.g., DB-5, Ultra-2) are usually the first choice of most experienced chromatographers. In cases where separation on that column is inadequate, all factors that influence component separation (lower temperatures or lower program rates, gas flow set at $\bar{u}_{opt}$ for the average partition ratios of the most critical solutes; Chapter 5) should be carefully optimized for the solutes in question. Only after the selectivity of the column under these optimized conditions has been proved inadequate is it advisable to consider using a more selective stationary phase. Increased selectivity often requires a stationary phase of greater polarity, and column efficiencies, high- and low-temperature tolerances, and general longevities usually vary inversely with the polarity of the stationary phase.

6.3   Stationary Phase Selectivity

In the early days of gas chromatography, judgments concerning the most appropriate stationary phase were entirely empirical and based largely on trial and error; columns were packed with almost anything from detergent chemicals to ill-defined lipid materials derived from various plant and animal sources. The

---

*Some forms of PEG are exceptions to this generality, in that they can yield very high efficiency columns.

occasional successes from these hit-or-miss efforts resulted in a bewildering array of poorly defined materials used as stationary phases, some of which endure to this day (e.g., the Apiezons and Carbowaxes, Ethofat, Fluorolube, the Igepals, THEED, Tergitol).

Selectivity decisions today are usually better formulated and based on qualitative judgments that try to relate the dispersive, acid–base, and dipole potentials of the solutes with those of specific stationary phases (Chapter 4). The hydrogen-bonding potential of alcohols, for example, would be exploited by the stationary phases in Table 4.1 in the order

$$PEG > cyanopropyl > trifluoropropyl = methyl\ silicone$$

whereas the trifluoropropyl silicone would exhibit improved selectivity for solutes that might be better differentiated on the basis of their dipole moments (including some enantiomers). The key word here is "differentiation"; although the hydrogen-bonding potential of the stationary phase would be beneficial in differentiating an alcohol from a hydrocarbon of similar boiling point, it would not help in differentiating 2-methyl-butanol from 3-methylbutanol, whose hydrogen-bonding characteristics would be quite similar.

The most elegant approach to stationary phase selectivity is based on predictions of the precise functionality that would be required to maximize the relative retentions of all solutes in a given mixture, followed by the synthesis of that stationary phase if it does not already exist (e.g., [4]). Alternatively, dissimilar columns can be serially coupled to achieve this goal (e.g. [5–7]). These methods are discussed in Chapter 8.

6.4   Selecting the Column Length

The most logical point of departure in selecting the column length is based on how much separation power, in terms of theoretical plates, will be required of that column. The decision concering (1) which stationary phase should be used and later decisions related to (2) column diameter and (3) stationary phase film thickness will also play roles here:

1. Stationary phase. Apolar columns usually exhibit lower values of h_{min} and are therefore capable of generating more theoretical plates per meter than columns containing more polar stationary phases (DB-Wax is an exception to this generalization).

2. Column diameter. If all other factors are equal, h_{min} varies directly with the diameter of the column. The disadvantages of smaller-diameter columns must be weighed against their greater separation potential per unit length and are discussed in Section 6.5.

3. Stationary phase film thickness. The positive aspects of this were considered in connection with Table 5.1; the negative aspects are normally of little

concern until d_f exceeds 0.4 μm and are also discussed in Chapter 5 and in Section 6.5.

If the column is destined for a specific analysis, the length can be selected to achieve the desired degree of resolution of the pair of solutes that is most challenging in the mixture. To utilize this route, some basic information will be required, e.g., the relative retention of those solutes (which is a function of the stationary phase and the temperature), the partition ratio of the more retained solute of the pair (which can be regulated to some degree by manipulating the column temperature and the thickness of the stationary phase film, Table 5.1), and the degree to which those solutes must be resolved (Fig. 1.5). Equation (1.22) can then be employed to estimate n_{req}.

For an all-round general-purpose column, 30 m is usually the most useful length. Shorter columns are frequently adequate for a given analysis, and because u_{opt}, OPGV, and the tightness of the van Deemter curve all vary inversely with column length (Chapter 4), short columns should be used where possible.

6.5 Selecting the Column Diameter

The 530-μm column is most valuable to those first converting from packed to open tubular columns because it permits the transition to be accomplished by one simple step at a time rather than one precipitous leap of total commitment. These large diameter open tubular columns still surpass packed columns in yielding improved component separation at higher sensitivities in shorter analysis times. Where maximum theoretical plate numbers are not required, the high-capacity large-diameter column also yields dividends in improved quantitation, both because of its superior inertness and because it accepts the total injected sample without fractionation. The ruggedness of these columns is such that they can be used by unskilled technicians for process line analysis on the production floor.

Where true capillary chromatography is desired, a smaller-diameter column will be required. In general, theoretical plate numbers vary inversely with column diameter, and the smaller the diameter of the column, the shorter the length of column that will be required for a given degree of separation. The most widely used diameters in open tubular gas chromatography are 320 and 250 μm, probably because they combine reasonably high efficiencies with reasonable sample loading capacities. Columns of 200- and 150-μm diameters are also available, and although they require a little more care, they can be very useful in some applications. As the column diameter decreases into the 100- and 50-μm region, the propensity for sample overloading and the difficulties of interfacing the column with the injector and the detector increase. Nevertheless, highly skilled chromatographers can use both of these small-diameter columns to good advantage in selected cases.

6.6 Selecting the Stationary Phase Film Thickness

The advent of securely bonded phases has made available a wide range of stationary phase film thicknesses. For general-purpose use, a standard film thickness (0.2–0.35 μm) is usually preferred because both column efficiency and sample carrying capacity are reasonable, and partition ratios of a wide range of solutes can be adjusted (via temperature) to yield good separation and high sensitivity. Thin-film columns ($d_f = 0.1$ μm) offer utility for very high boiling solutes, but such columns are more easily overloaded and are in many cases less inert. Film thicknesses ≥ 1 μm result in a significant increase in capacity and offer advantages in cold trapping of headspace volatiles [8,9]. Partition ratios become larger, which mandates either higher operating temperatures or a loss in sensitivity (Chapters 1 and 4). Much thicker film columns (up to 8 μm) have recently become available. Although these show great utility for the analysis of very low boiling solutes, they should never be regarded as general-purpose columns. Because they sacrifice one of the major advantages of open tubular columns [10], theoretical plate numbers are usually significantly lower, and for intermediate- or higher-k solutes the negative effects on separation, analysis times, and sensitivities can be substantial. Stationary phase diffusivity becomes critically important with these very thick film columns; in general, cyanopropyl phases are characterized by lower diffusivities and their use as thick films should be avoided [10,11].*

6.7 Column Installation

This section envisions installation in an open tubular-compatible instrument that has previously been used for that purpose; conversion of the packed column instrument to this capability, inlet and detector adaptation, and the desirability and selection of a makeup gas are all considered in Chapter 7.

The fused silica column should be installed in a configuration that does not subject the column to a high degree of strain; strained areas are more subject to surface corrosion, static fatigue, and breakage. The importance of avoiding strained mounting configurations increases with column diameter, because the degree of strain is directly proportional to the diameter of the tubing for any given coiling diameter. It is also important that the column support hardware contact the column support cage and not the column itself. Most commercial support cages are electropolished to minimize abrasion of the polyimide coating. In cases where the column comes in contact with other support hardware, movement encouraged by air currents within the oven may lead to abrasion of the polyimide coating and breakage will occur at those points. The very low thermal

*With high-diffusivity phases, the negative effects of thicker films are evidenced at $d_f \geq 0.4$ μm.

mass of the fused silica column means that it will respond "instantaneously" to radiant heat; there should not be a "line of sight" between the oven heater and the column. On instruments where the column can "see" the heating element, an aluminum foil barrier, which can be suspended or draped over the column support cage, will prevent the peak splitting and "Christmas tree" effects (see Chapter 10) that can be caused by exposure to sporadic radiant heat.

After freeing approximately 30 cm of column from each end of the support cage, the cage should be hung on the supporting hardware with the freed ends oriented toward the inlet and detector (Fig. 6.1). Ascertain the length required for one end to reach the inlet in an unstrained configuration, with the column terminating at the proper point in the inlet (see instructions of the inlet manufacturer or Chapter 3). In some cases, the distance the column should extend into the inlet beyond the base of the securing nut is specified. The proposed point of attachment (base of the ferrule) can be marked with typewriter correction fluid; when the column is held in that position, there should be no possibility of the column coming in contact with the oven wall. The column should now be cut at a point where it will be 2–4 cm too long.

To cut the column, use a diamond file or a carbide or diamond pencil (a very sharp razor blade can be used less satisfactorily), and scribe lightly through the polyimide; it is not necessary to scratch the silicia itself. Grasping the column on both sides of and as close to the scribe mark as possible, increase tension at that

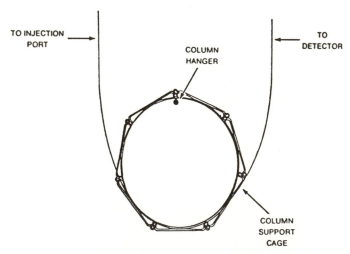

Fig. 6.1. Column configuration within the oven. Ideally, the column-supporting hardware should come in contact with the column cage and not the column per se, and the injector and detector arms of the column should curve smoothly to their respective connection points, subjecting the column to a minimum of strain. (Reprinted with permission of the copyright holder, J&W Scientific, Inc.)

point by pulling and bending; the column should snap cleanly. Especially with the larger-diameter open tubular columns, harsher cutting methods (e.g., snipping with scissors) can result in lines of fracture that extend longitudinally through an appreciable length of the tubing and lead to repetitive breakage.

A mounting nut and ferrule should now be slipped onto the column. For normal use, graphite ferrules are preferred; for some special applications, graphitized Vespel (trademark, DuPont) is required (see below). Vespel ferrules are denser and less permeable, but because they do not compress as readily as the graphite ferrule, their tolerance for discrepancies between the diameter of the column and the size of the hole in the ferrule is quite small. The excess length (2–4 cm) can now be trimmed from the end of the column; this procedure eliminates the possibility of depositing graphite scrapings inside the column. After examining the end of the column (at 10–20 × magnification) to verify a clean, square cut and the absence of chips of fused silica or polyimide, it is wise to activate the flow of carrier gas to flush from the inlet air that would otherwise be forced through the column. The inlet end can be secured while the carrier is still flowing. With graphite ferrules the nut should normally be tightened only one-half turn beyond the finger-tight position.

At this point it is advisable to insert the still unsecured detector end of the column into a small beaker of solvent and verify that the carrier is flowing through the column; a flow of approximately 2 cm^3/min is desirable. This flow can be continued during the rest of the installation procedure and should be continued for at least 30–40 min before energizing the column oven.

The length of column necessary to reach to the detector in a smooth and unstrained configuration, with the column end terminating at the proper point within the sensing zone (see below), should now be determined. Again, the column should be cut so that it is 2–4 cm too long, the securing nut and ferrule installed before the column is cut to the proper length, and the end examined carefully under magnification.

Several factors determine the ideal location of the end of the column within the detector. It is theoretically desirable to detect components at the instant they emerge from the column and to remove them immediately after detection; this would preserve the separation achieved by the column on closely eluting solutes. In practice, some degree of compromise is usually required. Columns that are positioned too close to the flame in a flame ionization detector (FID) sometimes suffer pyrolysis or carbonization of both the stationary phase and the outer polyimide coating on the final few millimeters of column; the pyrolyzed material can be extremely active, leading to reversible adsorption of some solutes (tailing peaks) and irreversible adsorption of others (total or partial component abstraction). Where possible, the column should be inserted completely through the jet, then pushed back so that it is flush with the jet orifice, and finally retracted an additional 2 mm. Retracting it farther than this can expose solutes to active sites, whose presence in both metal and quartz flame jets is readily demonstrable. Some older instruments utilize transfer lines to conduct the eluted solutes from the end of the column to the detector. Where possible, the potential activity of that configuration should be circumvented by extending the column completely through the transfer line. Other instruments incorporate sharp bends at the detector connection or restrictions at or within the flame jet that prohibit ideal positioning of the column end. Positioning is also critical with other types of detectors; the column should extend as far as possible into the detector to eliminate as many active sites as possible, but a loss in sensitivity results if it is positioned too deeply in the electron capture (EC), nitrogen/phosphorus (NP), or flame photometric (FP) detector or in the ion source of a mass spectrometer.

An electronic leak detector is the most satisfactory means of verifying the integrity of column connections. Where leak test solutions are employed, they are invariably aspirated into the system, where they contaminate the column, a problem that is exacerbated by surface activity in those solutions. Test solution residues that contaminate the column and/or the fittings can generate anomalous results, including baseline drift and high-sensitivity "noise." A column that has been leak-tested with one of the proprietary formulations may give satisfactory service with FID, but its use with NPD may be forever precluded. If a solution must be employed, it should be limited to volatile solvents that are compatible with the stationary phase and that produce a minimum detector response.

Before the oven is energized, the detector should be activated and a few microliters of an extremely volatile solute injected to verify that carrier gas is flowing through the column at a reasonable rate. With FID, methane (or propane/butane from a disposable cigarette lighter) is suitable; air, difluorodichloromethane (Freon 12; trademark, Dupont), or sulfur hexafluoride can be used with ECD; air can also be employed with GC/MS. Heating the column in the absence of carrier flow can seriously damage and may destroy the column. At this low temperature, the carrier gas velocity should be at least twice the value desired for $\bar{u}_{opt}$ at some higher temperature. This high flow also helps ensure that all air is flushed from both the inlet and the column before the temperature is increased. Unless the methane peak is needle-sharp, that problem should be corrected before proceeding further. The column securing nut at the detector should be loosened slightly, the column retracted another 1–2 mm, and the methane injection repeated. If the methane peak still remains broad or tails, the flow rate through the inlet (e.g., the split ratio) should be rechecked; the position, seal, and unbroken state of the inlet liner should be verified; and the position of the column in the inlet rechecked.

Most manufacturers supply a test chromatogram with the column, and it is wise to inject the same test sample under the manufacturer's test conditions, i.e., to duplicate the test chromatogram. It is not uncommon to find that an active solute exhibits severe tailing that was not demonstrated on the test chromatogram; this alerts the user to the fact that there are extracolumn sources of activity in the system, most probably in the inlet but possibly in the detector (see Chapter 10).

6.8   Column Conditioning

Packed columns normally require extensive high-temperature conditioning; open tubular columns rarely need the same degree of conditioning and may suffer damage when subjected to well-intentioned but ill-advised treatments.

Following installation and after all air is purged from the system, the column should be conditioned briefly at a temperature slightly above the maximum temperature at which it will be used. Exposure to higher temperatures places the column under stress and shortens the lifetime; such treatments should be avoided unless they serve some useful purpose. Samples that are run at moderate temperatures may also contain higher-boiling solutes that gradually accumulate on the column, and a periodic higher-temperature hold may be the easiest method to remove those residues from the column. All stationary phases are susceptible to stress from both oxygen and exposure to high temperatures. The two stresses may have a synergistic effect on the stationary phase: levels of oxygen that could be tolerated for several weeks at a lower temperature may destroy a column in a few minutes at higher temperatures. The oxygen tolerance of stationary phases in general decreases in the order

$$methyl\ polysiloxane \gg phenyl > cyanopropyl = trifluoropropyl \gg PEG$$

With oxygen-free carrier, the high-methyl polysiloxanes are usable at temperatures of 340–350°C, but even traces of oxygen can cause problems at temperatures above 200°C; C–Si bonds projecting from the polysiloxane backbone cleave to yield -OH (siloxane) groups [12]. Free silanol groups lead to a temperature-dependent degradation of the polysiloxane polymer [13–16] that will persist even in the absence of oxygen contamination. Methods for rectifying this problem are described in Chapter 10.

6.9   Optimizing Operational Parameters for Specific Columns

The graphs in Figs. 6.2–6.79, were generated to permit optimization of specific systems. In most cases, the analyst has a column on hand and wishes to optimize the operational conditions. Although that column has a certain length, there may be an advantage to employing a shorter segment of column. The column diameter, on the other hand, is usually fixed; for a given column, the stationary phase film thickness and the type of stationary phase are also fixed. The most appropriate curves are probably best selected by proceeding in the following order: (1) the diameter of the column to be used, (2) the length of that

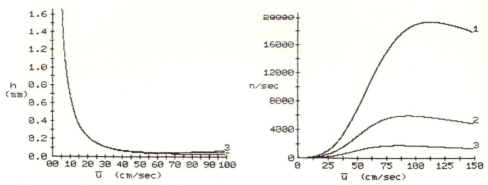

Fig. 6.2. Diameter 0.05 mm, length 5 m, d_f 0.2 μm, hydrogen carrier. Values assume ideality and are calculated for isothermal conditions. All values of h in millimeters, u_{opt} and OPGV in centimeters per second. Values for "time" are values of t_R for that solute at u_{opt} or OPGV.

| | | u_{opt} data | | | | OPGV data | | |
plot	k	h_{min}	u_{opt}	total n	time	OPGV	n/sec_{max}	total n	time
1	0.5	0.02	99	270,000	8	114	19283	127,000	7
2	2	0.04	73	125,000	21	95	5944	94,000	16
3	7	0.05	79	100,000	51	87	1820	83,700	46

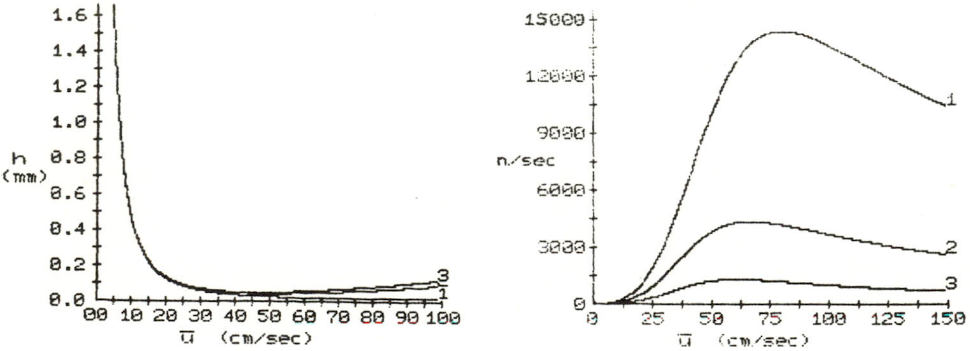

Fig. 6.3. Diameter 0.05 mm, length 10 m, d_f 0.2 μm, hydrogen carrier. Values assume ideality and are calculated for isothermal conditions. All values of h in millimeters, u_{opt} and OPGV in centimeters per second. Values for "time" are values of t_R for that solute at u_{opt} or OPGV.

| | | u_{opt} data | | | | OPGV data | | |
plot	k	h_{min}	u_{opt}	total n	time	OPGV	n/sec_{max}	total n	time
1	0.5	0.02	100	500,000	15	80	14400	270,000	19
2	2	0.04	56	250,000	54	67	4351	195,000	45
3	7	0.05	57	200,000	140	62	1308	169,000	129

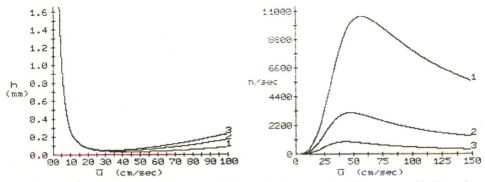

Fig. 6.4. Diameter 0.05 mm, length 20 m, d_f 0.2 μm, hydrogen carrier. Values assume ideality and are calculated for isothermal conditions. All values of h in millimeters, u_{opt} and OPGV in centimeter per second. Values for "time" are values of t_R for that solute at u_{opt} or OPGV.

		u_{opt} data			OPGV data			
plot	k	h_{min}	u_{opt}	total n	time	OPGV	n/sec$_{max}$	total n	time
1	0.5	0.02	56	1,000,000	54	57	10598	558,000	53
2	2	0.04	42	500,000	143	47	3159	403,000	128
3	7	0.05	41	400,000	390	44	937	341,000	364

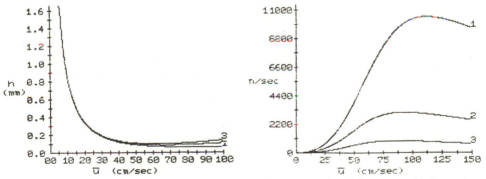

Fig. 6.5. Diameter 0.10 mm, length 10 m, d_f 0.4 μm, hydrogen carrier. Values assume ideality and are calculated for isothermal conditions. All values of h in millimeters, u_{opt} and OPGV in centimeters per second. Values for "time" are values of t_R for that solute at u_{opt} or OPGV.

		u_{opt} data			OPGV data			
plot	k	h_{min}	u_{opt}	total n	time	OPGV	n/sec$_{max}$	total n	time
1	0.5	0.06	100	167,000	15	113	10506	140,000	13
2	2	0.08	73	125,000	30	94	3120	100,000	32
3	7	0.10	76	100,000	80	85	924	87,000	94

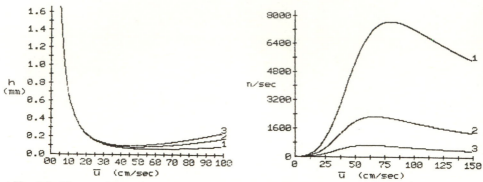

Fig. 6.6. Diameter 0.10 mm, length 20 m, d_f 0.4 μm, hydrogen carrier. Values assume ideality and are calculated for isothermal conditions. All values for h in millimeters, u_{opt} and OPGV in centimeters per second. Values for "time" are values of t_R for that solute at u_{opt} or OPGV.

			u_{opt} data			OPGV data			
plot	k	h_{min}	u_{opt}	total n	time	OPGV	n/sec_{max}	total n	time
1	0.5	0.06	74	333,000	41	81	7671	284,000	37
2	2	0.08	54	250,000	82	67	2257	202,000	74
3	7	0.10	55	200,000	216	62	662	171,000	198

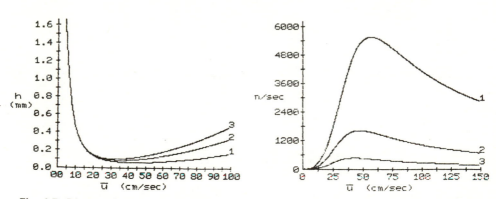

Fig. 6.7. Diameter 0.10 mm, length 40 m, d_f 0.4 μm, hydrogen carrier. Values assume ideality and are calculated for isothermal conditions. All values of h in millimeters, u_{opt} and OPGV in centimeters per second. Values for "time" are values of t_R for that solute at u_{opt} or OPGV.

			u_{opt} data			OPGV data			
plot	k	h_{min}	u_{opt}	total n	time	OPGV	n/sec_{max}	total n	time
1	0.5	0.06	54	667,000	111	57	5556	585,000	105
2	2	0.08	39	500,000	222	48	1623	406,000	250
3	7	0.10	39	400,000	593	43	474	353,000	744

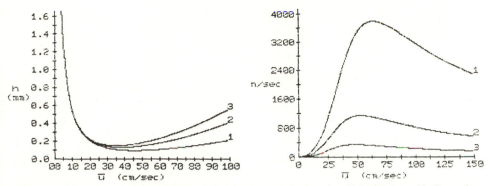

Fig. 6.8. Diameter 0.15 mm, length 15 m, d_f 0.4 μm, helium carrier. Values assume ideality and are calculated for isothermal conditions. All values of h in millimeters, u_{opt} and OPGV in centimeters per second. Values for "time" are values of t_R for that solute at u_{opt} or OPGV.

| | | u_{opt} data | | | | OPGV data | | |
plot	k	h_{min}	u_{opt}	total n	time	OPGV	n/sec_{max}	total n	time
1	0.5	0.09	49	167,000	46	64	3794	133,000	35
2	2	0.13	45	115,000	100	53	1148	97,000	85
3	7	0.15	41	100,000	293	49	346	85,000	245

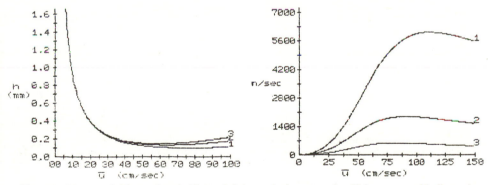

Fig. 6.9. Diameter 0.15 mm, length 15 m, d_f 0.4 μm, hydrogen carrier. Values assume ideality and are calculated for isothermal conditions. All values of h in millimeters, u_{opt} and OPGV in centimeters per second. Values for "time" are values of t_R for that solute at u_{opt} or OPGV.

| | | u_{opt} data | | | | OPGV data | | |
plot	k	h_{min}	u_{opt}	total n	time	OPGV	n/sec_{max}	total n	time
1	0.5	0.10	90	150,000	25	112	6026	121,000	20
2	2	0.13	69	115,000	65	93	1886	91,000	48
3	7	0.15	69	100,000	174	86	587	82,000	140

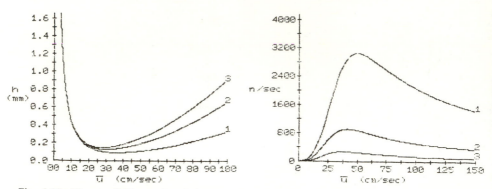

Fig. 6.10. Diameter 0.15 mm, length 25 m, d_f 0.4 μm, helium carrier. Values assume ideality and are calculated for isothermal conditions. All values of h in millimeters, u_{opt} and OPGV in centimeters per second. Values for "time" are values of t_R for that solute at u_{opt} or OPGV.

| | | | u_{opt} data | | | | OPGV data | | |
plot	k	h_{min}	u_{opt}	total n	time	OPGV	n/sec_{max}	total n	time
1	0.5	0.09	41	278,000	91	50	3047	229,000	75
2	2	0.13	36	192,000	208	41	913	167,000	183
3	7	0.15	33	167,000	606	38	273	144,000	526

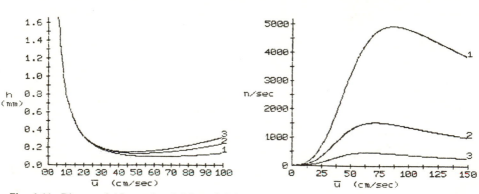

Fig. 6.11. Diameter 0.15 mm, length 25 m, d_f 0.4 μm, hydrogen carrier. Values assume ideality and are calculated for isothermal conditions. All values of h in millimeters, u_{opt} and OPGV in centimeters per second. Values for "time" are values of t_R for that solute at u_{opt} or OPGV.

| | | | u_{opt} data | | | | OPGV data | | |
plot	k	h_{min}	u_{opt}	total n	time	OPGV	n/sec_{max}	total n	time
1	0.5	0.10	76	250,000	49	87	4932	213,000	43
2	2	0.13	57	192,000	132	72	1520	158,000	104
3	7	0.15	55	167,000	363	67	466	139,000	299

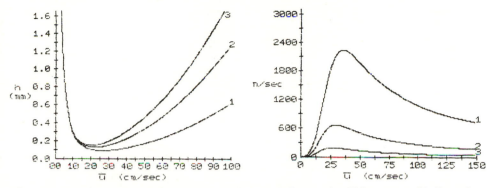

Fig. 6.12. Diameter 0.15 mm, length 50 m, d_f 0.4 μm, helium carrier. Values assume ideality and are calculated for isothermal conditions. All values of h in millimeters, u_{opt} and OPGV in centimeters per second. Values for "time" are values of t_R for that solute at u_{opt} or OPGV.

		u_{opt} data				OPGV data		
plot	k	h_{min}	u_{opt}	total n	time	OPGV	n/sec_{max}	total n	time
1	0.5	0.09	31	556,000	242	35	2239	480,000	214
2	2	0.13	26	385,000	577	29	664	344,000	517
3	7	0.15	24	333,000	1667	28	196	280,000	1429

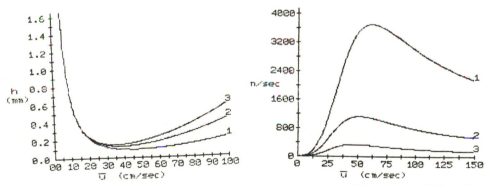

Fig. 6.13. Diameter 0.15 mm, length 50 m, d_f 0.4 μm, hydrogen carrier. Values assume ideality and are calculated for isothermal conditions. All values of h in millimeters, u_{opt} and OPGV in centimeters per second. Values for "time" are values of t_R for that solute at u_{opt} or OPGV.

		u_{opt} data				OPGV data		
plot	k	h_{min}	u_{opt}	total n	time	OPGV	n/sec_{max}	total n	time
1	0.5	0.09	47	556,000	160	62	3696	447,000	121
2	2	0.13	43	385,000	349	52	1119	323,000	288
3	7	0.15	40	333,000	1000	48	337	281,000	833

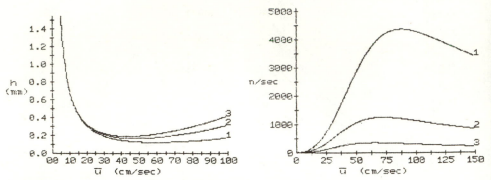

Fig. 6.14. Diameter 0.20 mm, length 10 m, d_f 0.4 μm, helium carrier. Values assume ideality and are calculated for isothermal conditions. All values of h in millimeters, u_{opt} and OPGV in centimeters per second. Values for "time" are values of t_R for that solute at u_{opt} or OPGV.

		u_{opt} data				OPGV data		
plot	k	h_{min}	u_{opt}	total n	time	OPGV	n/sec$_{max}$	total n	time
1	0.5	0.12	68	83,300	22	89	4361	73,500	17
2	2	0.17	59	58,800	51	72	1251	52,100	42
3	7	0.19	50	52,600	160	68	360	43,400	118

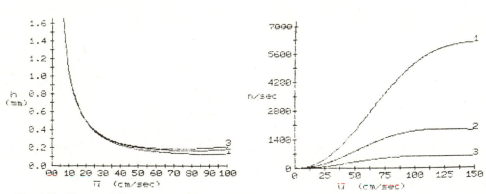

Fig. 6.15. Diameter 0.20 mm, length 10 m, d_f 0.4 μm, hydrogen carrier. Values assume ideality and are calculated for isothermal conditions. All values of h in millimeters, u_{opt} and OPGV in centimeters per second. Values for "time" are values of t_R for that solute at u_{opt} or OPGV.

		u_{opt} data				OPGV data		
plot	k	h_{min}	u_{opt}	total n	time	OPGV	n/sec$_{max}$	total n	time
1	0.5	0.12	100	83,300	15	150	6300	63,000	10
2	2	0.17	89	58,800	34	125	1897	45,500	24
3	7	0.19	82	52,600	98	118	585	39,700	68

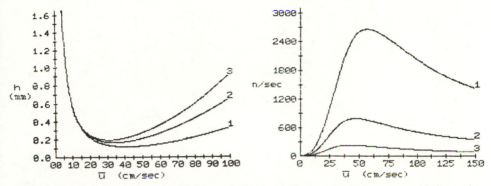

Fig. 6.16. Diameter 0.20 mm, length 25 m, d_f 0.4 μm, helium carrier. Values assume ideality and are calculated for isothermal conditions. All values of h in millimeters, u_{opt} and OPGV in centimeters per second. Values for "time" are values of t_R for that solute at u_{opt} or OPGV.

		u_{opt} data				OPGV data		
plot	k	h_{min}	u_{opt}	total n	time	OPGV	n/sec_{max}	total n	time
1	0.5	0.12	48	208,000	78	57	2657	175,000	66
2	2	0.17	40	147,000	188	48	788	123,000	156
3	7	0.19	33	132,000	606	43	234	109,000	465

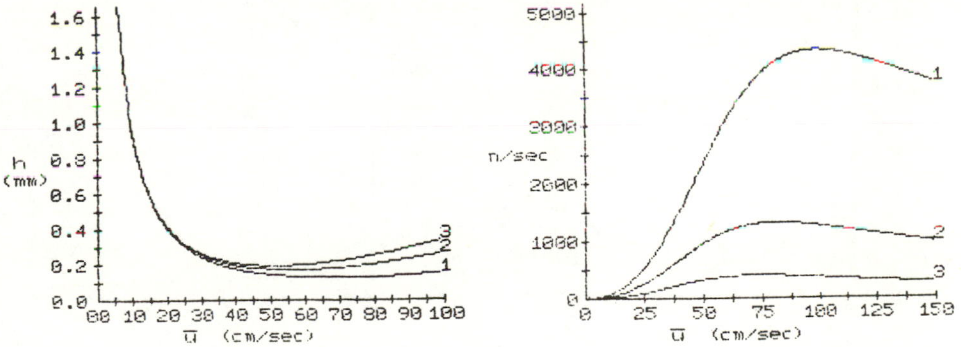

Fig. 6.17. Diameter 0.20 mm, length 25 m, d_f 0.4 μm, hydrogen carrier. Values assume ideality and are calculated for isothermal conditions. All values of h in millimeters, u_{opt} and OPGV in centimeters per second. Values for "time" are values of t_R for that solute at u_{opt} or OPGV.

		u_{opt} data				OPGV data		
plot	k	h_{min}	u_{opt}	total n	time	OPGV	n/sec_{max}	total n	time
1	0.5	0.13	85	192,000	44	99	4330	164,000	38
2	2	0.17	64	147,000	117	82	1315	120,000	91
3	7	0.19	56	132,000	357	75	398	106,000	267

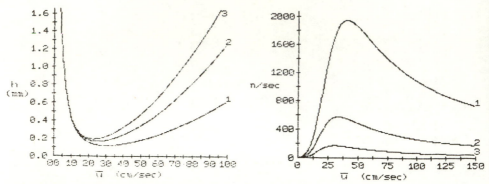

Fig. 6.18. Diameter 0.20 mm, length 50 m, d_f 0.4 μm, helium carrier. Values assume ideality and are calculated for isothermal conditions. All values of h in millimeters, u_{opt} and OPGV in centimeters per second. Values for "time" are values of t_R for that solute at u_{opt} or OPGV.

| | | | u_{opt} data | | | OPGV data | | |
plot	k	h_{min}	u_{opt}	total n	time	OPGV	n/sec_{max}	total n	time
1	0.5	0.12	36	417,000	208	41	1950	357,000	183
2	2	0.17	29	294,000	517	35	574	246,000	429
3	7	0.19	23	263,000	1739	32	169	211,000	1250

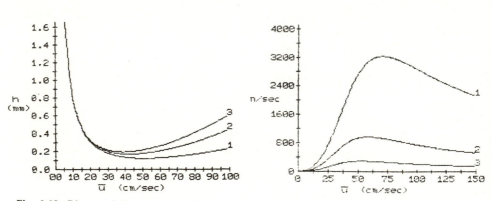

Fig. 6.19. Diameter 0.20 mm, length 50 m, d_f 0.4 μm, hydrogen carrier. Values assume ideality and are calculated for isothermal conditions. All values of h in millimeters, u_{opt} and OPGV in centimeters per second. Values for "time" are values of t_R for that solute at u_{opt} or OPGV.

| | | | u_{opt} data | | | OPGV data | | |
plot	k	h_{min}	u_{opt}	total n	time	OPGV	n/sec_{max}	total n	time
1	0.5	0.12	57	417,000	132	71	3240	342,000	106
2	2	0.17	48	294,000	313	59	971	247,000	254
3	7	0.19	39	263,000	1026	55	290	211,000	727

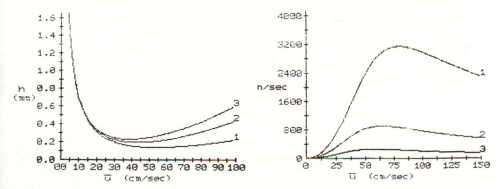

Fig. 6.20. Diameter 0.25 mm, length 15 m, d_f 0.25 μm, helium carrier. Values assume ideality and are calculated for isothermal conditions. All values of h in millimeters, u_{opt} and OPGV in centimeters per second. Values for "time" are values of t_R for that solute at u_{opt} or OPGV.

		u_{opt} data					OPGV data		
plot	k	h_{min}	u_{opt}	total n	time	OPGV	n/sec_{max}	total n	time
1	0.5	0.14	65	107,000	35	80	3141	88,000	28
2	2	0.20	52	75,000	87	66	899	61,000	68
3	7	0.23	44	65,000	273	61	259	51,000	197

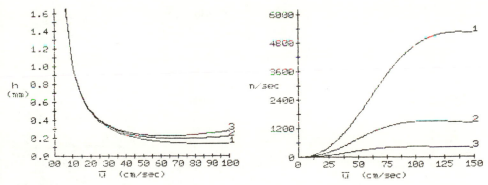

Fig. 6.21. Diameter 0.25 mm, length 15 m, d_f 0.25 μm, hydrogen carrier. Values assume ideality and are calculated for isothermal conditions. All values of h in millimeters, u_{opt} and OPGV in centimeters per second. Values for "time" are values of t_R for that solute at u_{opt} or OPGV.

		u_{opt} data					OPGV data		
plot	k	h_{min}	u_{opt}	total n	time	OPGV	n/sec_{max}	total n	time
1	0.5	0.14	99	107,000	23	139	5234	85,000	16
2	2	0.20	86	75,000	53	114	1504	59,000	39
3	7	0.23	74	65,000	162	105	435	48,000	114

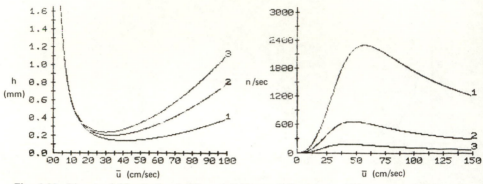

Fig. 6.22. Diameter 0.25 mm, length 30 m, d_f 0.25 μm, helium carrier. Values assume ideality and are calculated for isothermal conditions. All values of h in millimeters, u_{opt} and OPGV in centimeters per second. Values for "time" are values of t_R for that solute at u_{opt} or OPGV.

			u_{opt} data			OPGV data		
plot	k	h_{min}	u_{opt}	total n	time	OPGV	n/sec_{max}	total n	time
1	0.5	0.14	48	214,000	94	58	2301	179,000	78
2	2	0.20	38	150,000	237	48	662	124,000	188
3	7	0.23	31	130,000	774	45	191	102,000	533

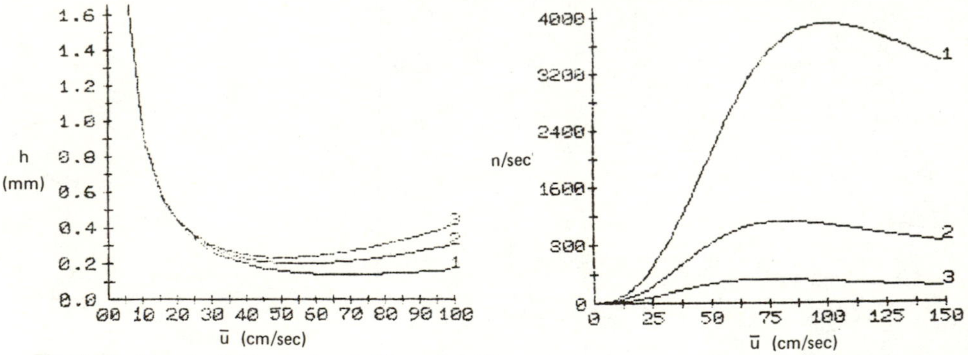

Fig. 6.23. Diameter 0.25 mm, length 30 m, d_f 0.25 μm, hydrogen carrier. Values assume ideality and are calculated for isothermal conditions. All values of h in millimeters, u_{opt} and OPGV in centimeters per second. Values for "time" are values of t_R for that solute at u_{opt} or OPGV.

			u_{opt} data			OPGV data		
plot	k	h_{min}	u_{opt}	total n	time	OPGV	n/sec_{max}	total n	time
1	0.5	0.14	79	214,000	57	101	3902	174,000	45
2	2	0.20	65	150,000	138	83	1126	122,000	108
3	7	0.23	54	130,000	444	78	326	100,000	308

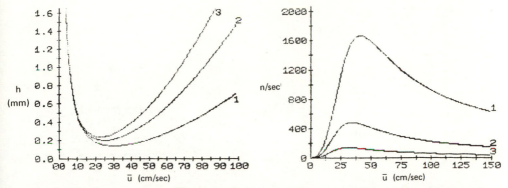

Fig. 6.24. Diameter 0.25 mm, length 60 m, d_f 0.25 μm, helium carrier. Values assume ideality and are calculated for isothermal conditions. All values of h in millimeters, u_{opt} and OPGV in centimeters per second. Values for "time" are values of t_R for that solute at u_{opt} or OPGV.

			u_{opt} data				OPGV data		
plot	k	h_{min}	u_{opt}	total n	time	OPGV	n/sec_{max}	total n	time
1	0.5	0.14	34	429,000	265	41	1665	366,000	220
2	2	0.20	27	300,000	667	34	480	254,000	529
3	7	0.23	25	250,000	1920	32	139	209,000	1500

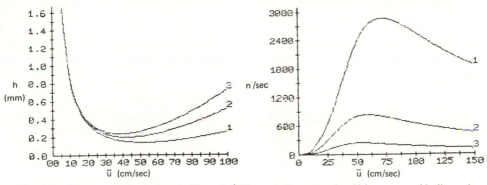

Fig. 6.25. Diameter 0.25 mm, length 60 m, d_f 0.25 μm, hydrogen carrier. Values assume ideality and are calculated for isothermal conditions. All values of h in millimeters, u_{opt} and OPGV in centimeters per second. Values for "time" are values of t_R for that solute at u_{opt} or OPGV.

			u_{opt} data				OPGV data		
plot	k	h_{min}	u_{opt}	total n	time	OPGV	n/sec_{max}	total n	time
1	0.5	0.14	58	429,000	155	72	2854	357,000	125
2	2	0.20	46	300,000	391	59	825	252,000	305
3	7	0.23	44	250,000	1091	56	239	205,000	857

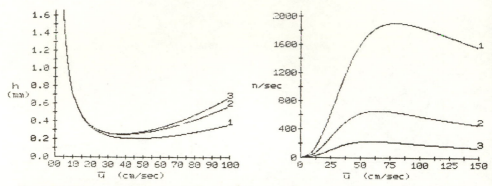

Fig. 6.26. Diameter 0.25 mm, length 15 m, d_f 1.0 μm, helium carrier. Values assume ideality and are calculated for isothermal conditions. All values of h in millimeters, u_{opt} and OPGV in centimeters per second. Values for "time" are values of t_R for that solute at u_{opt} or OPGV.

		$\overline{u}_{opt}$ data				OPGV data		
plot	k	h_{min}	u_{opt}	total n	time	OPGV	n/sec_{max}	total n	time
1	0.5	0.21	55	71,000	41	81	1899	53,000	28
2	2	0.25	44	60,000	102	66	655	45,000	68
3	7	0.26	42	58,000	286	62	227	44,000	194

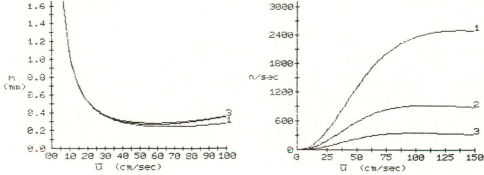

Fig. 6.27. Diameter 0.25 mm, length 15 m, d_f 1.0 μm, hydrogen carrier. Values assume ideality and are calculated for isothermal conditions. All values of h in millimeters, u_{opt} and OPGV in centimeters per second. Values for "time" are values of t_R for that solute at u_{opt} or OPGV.

		u_{opt} data				OPGV data		
plot	k	h_{min}	u_{opt}	total n	time	OPGV	n/sec_{max}	total n	time
1	0.5	0.25	82	60,000	27	138	2505	40,800	16
2	2	0.28	61	53,600	74	116	925	35,900	39
3	7	0.27	67	55,600	179	105	352	40,200	114

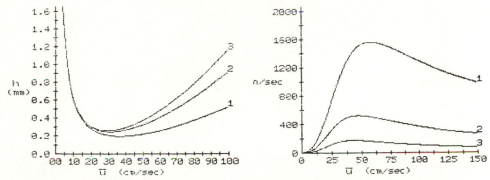

Fig. 6.28. Diameter 0.25 mm, length 30, d_f 1.0 μm, helium carrier. Values assume ideality and are calculated for isothermal conditions. All values of h in millimeters, u_{opt} and OPGV in centimeters per second. Values for "time" are values of t_R for that solute at u_{opt} or OPGV.

		u_{opt} data				OPGV data		
plot	k	h_{min}	u_{opt}	total n	time	OPGV	n/sec_{max}	total n	time
1	0.5	0.19	39	158,000	115	58	1556	121,000	78
2	2	0.24	33	125,000	272	48	519	97,000	188
3	7	0.26	34	115,000	706	45	173	92,000	533

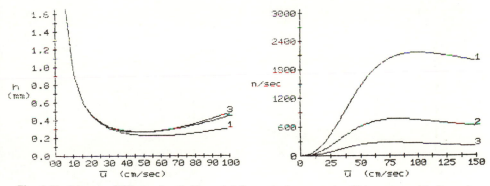

Fig. 6.29. Diameter 0.25 mm, length 30 m, d_f 1.0 μm, hydrogen carrier. Values assume ideality and are calculated for isothermal conditions. All values of h in millimeters, u_{opt} and OPGV in centimeters per second. Values for "time" are values of t_R for that solute at u_{opt} or OPGV.

		u_{opt} data				OPGV data		
plot	k	h_{min}	u_{opt}	total n	time	OPGV	n/sec_{max}	total n	time
1	0.5	0.23	67	130,000	67	102	2153	95,000	44
2	2	0.27	55	111,000	164	84	767	82,000	107
3	7	0.27	56	111,000	429	78	277	85,000	308

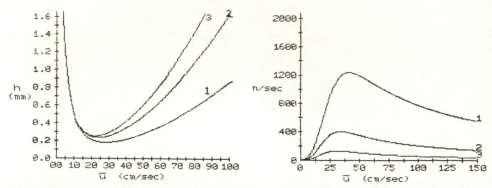

Fig. 6.30. Diameter 0.25 mm, length 60 m, d_f 1.0 μm, helium carrier. Values assume ideality and are calculated for isothermal conditions. All values of h in millimeters, u_{opt} and OPGV in centimeters per second. Values for "time" are values of t_R for that solute at u_{opt} or OPGV.

		u_{opt} data				OPGV data			
plot	k	h_{min}	u_{opt}	total n	time	OPGV	n/sec_{max}	total n	time
1	0.5	0.18	31	333,000	290	42	1236	265,000	214
2	2	0.24	28	250,000	643	35	400	206,000	514
3	7	0.25	23	240,000	2087	33	129	188,000	1455

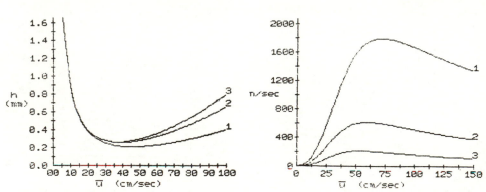

Fig. 6.31. Diameter 0.25 mm, length 60 m, d_f 1.0 μm, hydrogen carrier. Values assume ideality and are calculated for isothermal conditions. All values of h in millimeters, u_{opt} and OPGV in centimeters per second. Values for "time" are values of t_R for that solute at u_{opt} or OPGV.

		u_{opt} data				OPGV data			
plot	k	h_{min}	u_{opt}	total n	time	OPGV	n/sec_{max}	total n	time
1	0.5	0.21	52	286,000	173	73	1790	221,000	123
2	2	0.26	45	231,000	400	61	614	181,000	295
3	7	0.26	40	231,000	1200	55	212	185,000	873

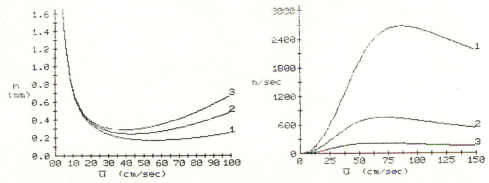

Fig. 6.32. Diameter 0.32 mm, length 15 m, d_f 0.25 μm, helium carrier. Values assume ideality and are calculated for isothermal conditions. All values of h in millimeters, u_{opt} and OPGV in centimeters per second. Values for "time" are values of t_R for that solute at u_{opt} or OPGV.

			u_{opt} data			OPGV data		
plot	k	h_{min}	u_{opt}	total n	time	OPGV	n/sec_{max}	total n	time
1	0.5	0.17	62	88,200	36	87	2674	69,200	26
2	2	0.24	45	62,500	100	72	751	47,000	62
3	7	0.29	44	51,700	273	66	214	38,900	182

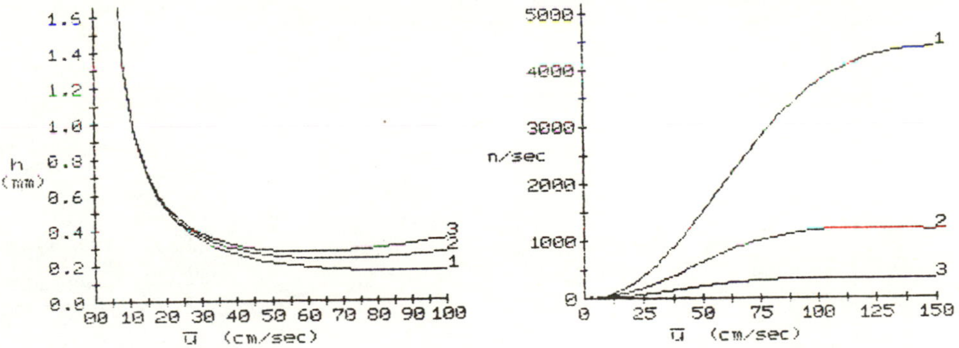

Fig. 6.33. Diameter 0.32 mm, length 15 m, d_f 0.25 μm, hydrogen carrier. Values assume ideality and are calculated for isothermal conditions. All values of h in millimeters, u_{opt} and OPGV in centimeters per second. Values for "time" are values of t_R for that solute at u_{opt} or OPGV.

			u_{opt} data			OPGV data		
plot	k	h_{min}	u_{opt}	total n	time	OPGV	n/sec_{max}	total n	time
1	0.5	0.17	100	88,200	22	155	2674	38,800	15
2	2	0.24	83	62,500	54	124	1230	44,600	36
3	7	0.28	71	53,600	169	115	350	36,500	104

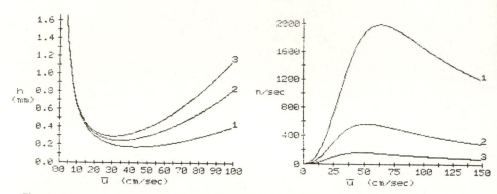

Fig. 6.34. Diameter 0.32 mm, length 30, d_f 0.25 μm, helium carrier. Values assume ideality and are calculated for isothermal conditions. All values of h in millimeters, u_{opt} and OPGV in centimeters per second. Values for "time" are values of t_R for that solute at u_{opt} or OPGV.

		u_{opt} data				OPGV data		
plot	k	h_{min}	u_{opt}	total n	time	OPGV	n/sec_{max}	total n	time
1	0.5	0.17	45	177,000	100	64	1995	140,000	70
2	2	0.25	40	120,000	225	53	567	96,000	170
3	7	0.29	33	104,000	727	48	163	81,500	500

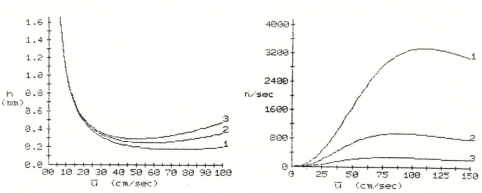

Fig. 6.35. Diameter 0.32 mm, length 30 m, d_f 0.25 μm, hydrogen carrier. Values assume ideality and are calculated for isothermal conditions. All values of h in millimeters, u_{opt} and OPGV in centimeters per second. Values for "time" are values of t_R for that solute at u_{opt} or OPGV.

		u_{opt} data				OPGV data		
plot	k	h_{min}	u_{opt}	total n	time	OPGV	n/sec_{max}	total n	time
1	0.5	0.18	88	167,000	51	110	3356	137,000	41
2	2	0.25	66	120,000	136	91	952	94,000	99
3	7	0.29	55	103,000	436	84	273	78,000	286

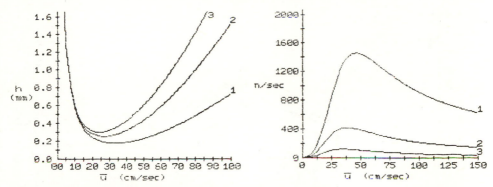

Fig. 6.36. Diameter 0.32 mm, length 60 m, d_f 0.25 μm, helium carrier. Values assume ideality and are calculated for isothermal conditions. All values of h in millimeters, u_{opt} and OPGV in centimeters per second. Values for "time" are values of t_R for that solute at u_{opt} or OPGV.

		u_{opt} data				OPGV data			
plot	k	h_{min}	u_{opt}	total n	time	OPGV	n/sec_{max}	total n	time
1	0.5	0.18	39	333,000	230	47	1457	280,000	192
2	2	0.25	28	240,000	642	39	417	193,000	462
3	7	0.30	26	200,000	1846	37	120	156,000	1297

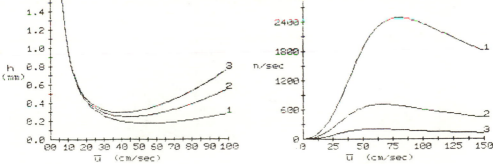

Fig. 6.37. Diameter 0.32 mm, length 60 m, d_f 0.25 μm, hydrogen carrier. Values assume ideality and are calculated for isothermal conditions. All values of h in millimeters. u_{opt} and OPGV in centimeters per second. Values for "time" are values of t_R for that solute at u_{opt} or OPGV.

		u_{opt} data				OPGV data			
plot	k	h_{min}	u_{opt}	total n	time	OPGV	n/sec_{max}	total n	time
1	0.5	0.18	65	333,000	138	80	2489	280,000	113
2	2	0.25	49	240,000	367	66	712	194,000	273
3	7	0.30	46	200,000	1043	62	205	159,000	774

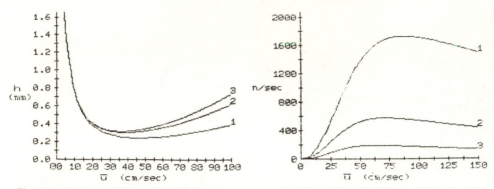

Fig. 6.38. Diameter 0.32 mm, length 15 m, d_f 1.0 μm, helium carrier. Values assume ideality and are calculated for isothermal conditions. All values of h in millimeters, u_{opt} and OPGV in centimeters per second. Values for "time" are values of t_R for that solute at u_{opt} or OPGV.

| | | u_{opt} data | | | | OPGV data | | |
plot	k	h_{min}	u_{opt}	total n	time	OPGV	n/sec_{max}	total n	time
1	0.5	0.24	51	62,500	44	88	1718	44,000	26
2	2	0.30	44	50,000	102	74	572	34,800	61
3	7	0.31	39	48,400	308	66	192	34,900	182

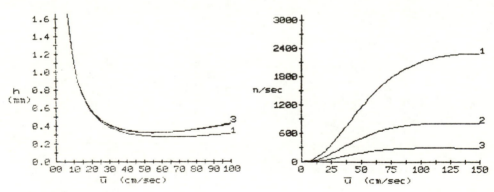

Fig. 6.39. Diameter 0.32 mm, length 15 m, d_f 1.0 μm, hydrogen carrier. Values assume ideality and are calculated for isothermal conditions. All values of h in millimeters, u_{opt} and OPGV in centimeters per second. Values for "time" are values of t_R for that solute at u_{opt} or OPGV.

| | | u_{opt} data | | | | OPGV data | | |
plot	k	h_{min}	u_{opt}	total n	time	OPGV	n/sec_{max}	total n	time
1	0.5	0.28	79	53,600	28	150	2260	33,900	15
2	2	0.33	65	45,500	36	125	814	29,300	36
3	7	0.32	54	46,900	222	117	294	30,200	103

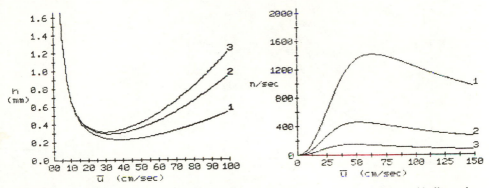

Fig. 6.40. Diameter 0.32 mm, length 30 m, d_f 1.0 μm, helium carrier. Values assume ideality and are calculated for isothermal conditions. All values of h in millimeters, u_{opt} and OPGV in centimeters per second. Values for "time" are values of t_R for that solute at u_{opt} or OPGV.

| | | u_{opt} data | | | | OPGV data | | |
plot	k	h_{min}	u_{opt}	total n	time	OPGV	n/sec$_{max}$	total n	time
1	0.5	0.23	43	130,000	105	64	1410	99,100	70
2	2	0.29	33	103,000	170	53	459	77,900	170
3	7	0.31	31	96,800	471	51	149	70,100	471

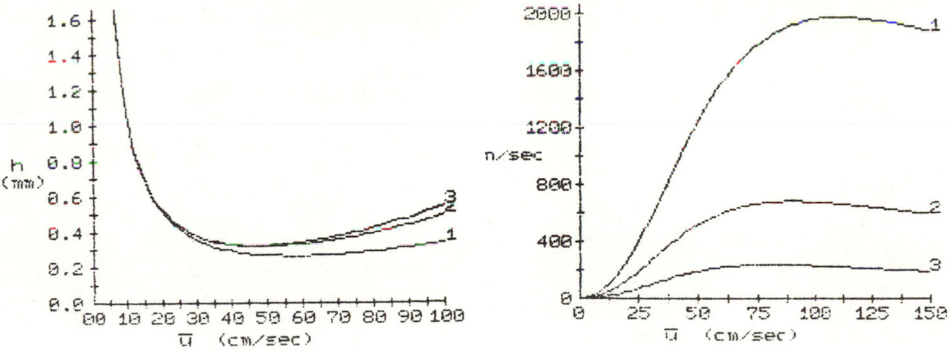

Fig. 6.41. Diameter 0.32 mm, length 30 m, d_f 1.0 μm, hydrogen carrier. Values assume ideality and are calculated for isothermal conditions. All values of h in millimeters, u_{opt} and OPGV in centimeters per second. Values for "time" are values of t_R for that solute at u_{opt} or OPGV.

| | | u_{opt} data | | | | OPGV data | | |
plot	k	h_{min}	u_{opt}	total n	time	OPGV	n/sec$_{max}$	total n	time
1	0.5	0.26	61	115,000	74	111	1976	80,100	41
2	2	0.32	55	93,800	164	92	682	66,700	98
3	7	0.32	50	93,800	480	83	238	68,800	289

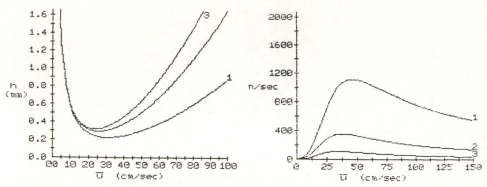

Fig. 6.42. Diameter 0.32 mm, length 60 m, d_f 1.0 μm, helium carrier. Values assume ideality and are calculated for isothermal conditions. All values of h in millimeters, u_{opt} and OPGV in centimeters per second. Values for "time" are values of t_R for that solute at u_{opt} or OPGV.

		u_{opt} data				OPGV data			
plot	k	h_{min}	u_{opt}	total n	time	OPGV	n/sec$_{max}$	total n	time
1	0.5	0.22	34	273,000	265	47	1118	214,000	191
2	2	0.29	28	207,000	643	38	356	169,000	474
3	7	0.32	27	188,000	1778	36	113	151,000	1333

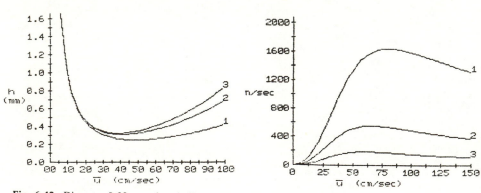

Fig. 6.43. Diameter 0.32 mm, length 60 m, d_f 1.0 μm, hydrogen carrier. Values assume ideality and are calculated for isothermal conditions. All values of h in millimeters, u_{opt} and OPGV in centimeters per second. Values for "time" are values of t_R for that solute at u_{opt} or OPGV.

		u_{opt} data				OPGV data			
plot	k	h_{min}	u_{opt}	total n	time	OPGV	n/sec$_{max}$	total n	time
1	0.5	0.25	55	240,000	164	80	1640	185,000	113
2	2	0.31	44	194,000	409	65	550	152,000	277
3	7	0.32	41	188,000	1171	60	185	148,000	800

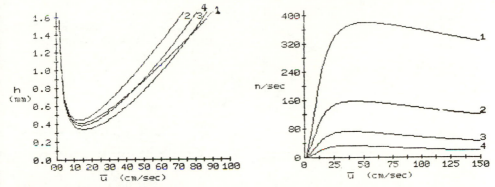

Fig. 6.44. Diameter 0.32 mm, length 15 m, d_f 3.0 μm, nitrogen carrier. Values assume ideality and are calculated for isothermal conditions. All values of h in millimeters, u_{opt} and OPGV in centimeters per second. Values for "time" are values of t_R for that solute at u_{opt} or OPGV.

| | | u_{opt} data | | | | OPGV data | | |
plot	k	h_{min}	u_{opt}	total n	time	OPGV	n/sec_{max}	total n	time
1	0.5	0.41	16	36,600	141	58	380	14,700	39
2	2	0.45	15	33,300	300	49	157	14,400	92
3	7	0.38	14	39,500	857	41	73	21,400	293
4	20	0.34	17	44,100	1853	46	33	22,600	685

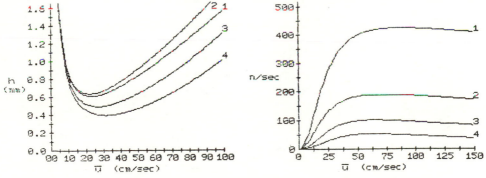

Fig. 6.45. Diameter 0.32 mm, length 15 m, d_f 3.0 μm, helium carrier. Values assume ideality and are calculated for isothermal conditions. All values of h in millimeters, u_{opt} and OPGV in centimeters per second. Values for "time" are values of t_R for that solute at u_{opt} or OPGV.

| | | u_{opt} data | | | | OPGV data | | |
plot	k	h_{min}	u_{opt}	total n	time	OPGV	n/sec_{max}	total n	time
1	0.5	0.61	26	24,600	87	91	425	10,500	25
2	2	0.63	23	23,800	196	83	189	10,200	54
3	7	0.49	28	30,600	429	71	102	17,200	169
4	20	0.39	32	38,500	984	71	53	23,500	444

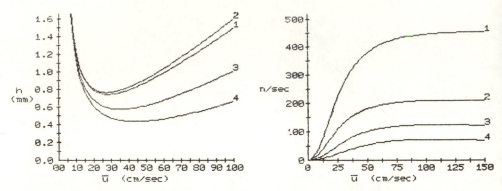

Fig. 6.46. Diameter 0.32 mm, length 15 m, d_f 3.0 μm, hydrogen carrier. Values assume ideality and are calculated for isothermal conditions. All values of h in millimeters, u_{opt} and OPGV in centimeters per second. Values for "time" are values of t_R for that solute at u_{opt} or OPGV.

			u_{opt} data				OPGV data		
plot	k	h_{min}	u_{opt}	total n	time	OPGV	n/sec$_{max}$	total n	time
1	0.5	0.74	29	20,300	78	91	425	10,500	25
2	2	0.76	27	19,700	167	139	210	6,800	32
3	7	0.58	40	25,900	300	120	125	12,500	100
4	20	0.44	51	34,100	618	119	73	19,300	265

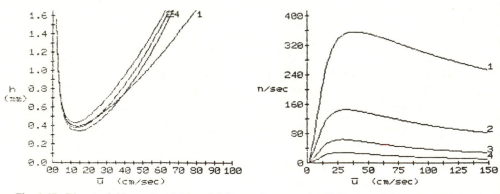

Fig. 6.47. Diameter 0.32 mm, length 30 m, d_f 3.0 μm, nitrogen carrier. Values assume ideality and are calculated for isothermal conditions. All values of h in millimeters, u_{opt} and OPGV in centimeters per second. Values for "time" are values of t_R for that solute at u_{opt} or OPGV.

			u_{opt} data				OPGV data		
plot	k	h_{min}	u_{opt}	total n	time	OPGV	n/sec$_{max}$	total n	time
1	0.5	0.39	14	77,000	321	42	356	38,100	107
2	2	0.43	13	69,800	692	35	144	37,000	257
3	7	0.38	15	78,900	1600	29	64	53,000	828
4	20	0.34	16	88,200	3938	34	28	51,900	1853

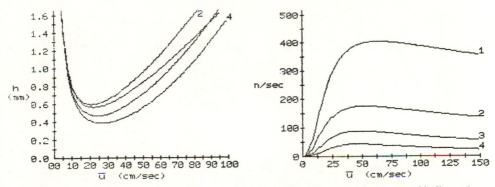

Fig. 6.48. Diameter 0.32 mm, length 30 m, d_f 3.0 µm, helium carrier. Values assume ideality and are calculated for isothermal conditions. All values of h in millimeters, u_{opt} and OPGV in centimeters per second. Values for "time" are values of t_R for that solute at u_{opt} or OPGV.

		u_{opt} data				OPGV data		
plot	k	h_{min}	u_{opt}	total n	time	OPGV	n/sec_{max}	total n	ti
1	0.5	0.56	22	53,600	205	68	403	26,700	66
2	2	0.59	21	50,800	429	58	175	27,200	155
3	7	0.47	27	63,800	889	50	89	42,700	480
4	20	0.39	31	76,900	2032	48	44	57,800	1313

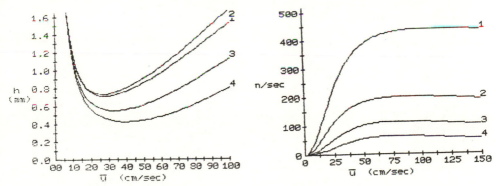

Fig. 6.49. Diameter 0.32 mm, length 30 m, d_f 3.0 µm, hydrogen carrier. Values assume ideality and are calculated for isothermal conditions. All values of h in millimeters, u_{opt} and OPGV in centimeters per second. Values for "time" are values of t_R for that solute at u_{opt} or OPGV.

		u_{opt} data				OPGV data		
plot	k	h_{min}	u_{opt}	total n	time	OPGV	n/sec_{max}	total n	time
1	0.5	0.71	30	42,300	150	118	439	16,700	38
2	2	0.73	29	41,100	310	100	200	18,000	90
3	7	0.55	36	54,500	667	94	113	28,900	255
4	20	0.42	41	71,400	1537	81	63	49,000	778

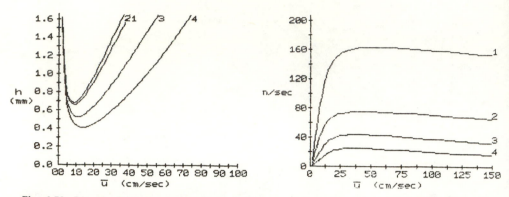

Fig. 6.50. Diameter 0.32 mm, length 15 m, d_f 5.0 μm, nitrogen carrier. Values assume ideality and are calculated for isothermal conditions. All values of h in millimeters, u_{opt} and OPGV in centimeters per second. Values for "time" are values of t_R for that solute at u_{opt} or OPGV.

		u_{opt} data				OPGV data			
plot	k	h_{min}	u_{opt}	total n	time	OPGV	n/sec_{max}	total n	time
1	0.5	0.66	9	22,700	250	71	162	5,100	32
2	2	0.68	8	22,100	563	55	75	6,100	82
3	7	0.53	12	28,300	1000	48	44	11,000	250
4	20	0.41	15	36,600	2100	50	25	15,800	630

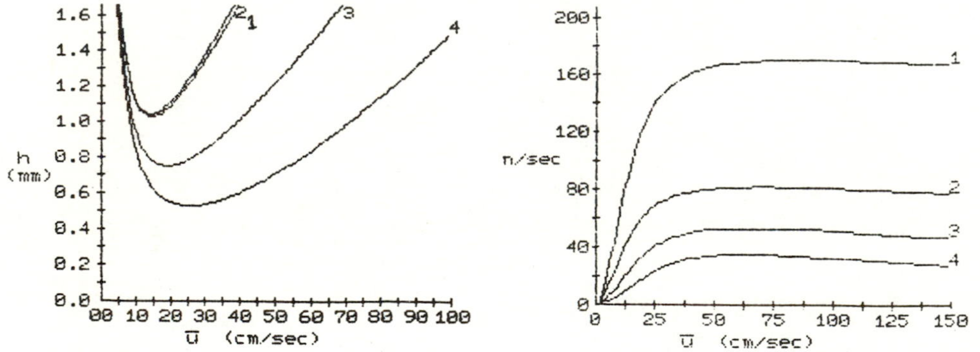

Fig. 6.51. Diameter 0.32 mm, length 15 m, d_f 5.0 μm, helium carrier. Values assume ideality and are calculated for isothermal conditions. All values of h in millimeters, u_{opt} and OPGV in centimeters per second. Values for "time" are values of t_R for that solute at u_{opt} or OPGV.

		u_{opt} data				OPGV data			
plot	k	h_{min}	u_{opt}	total n	time	OPGV	n/sec_{max}	total n	time
1	0.5	1.03	15	14,600	150	110	170	3,500	20
2	2	1.04	14	14,400	321	75	82	4,500	60
3	7	0.75	20	20,000	600	79	53	8,100	152
4	20	0.53	28	28,300	1125	77	35	14,300	409

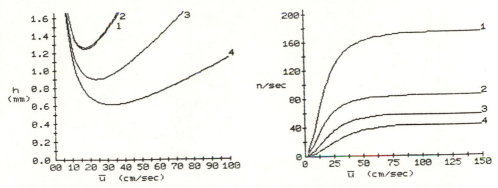

Fig. 6.52. Diameter 0.32 mm, length 15 m, d_f 5.0 μm, hydrogen carrier. Values assume ideality and are calculated for isothermal conditions. All values of h in millimeters, u_{opt} and OPGV in centimeters per second. Values for "time" are values of t_R for that solute at u_{opt} or OPGV.

| | | u_{opt} data | | | | OPGV data | | |
plot	k	h_{min}	u_{opt}	total n	time	OPGV	n/sec_{max}	total n	time
1	0.5	1.25	18	12,000	125	110	170	3,500	20
2	2	1.26	18	11,900	237	75	82	4,900	60
3	7	0.90	23	16,700	522	79	53	8,100	152
4	20	0.62	36	24,200	875	112	43	12,100	281

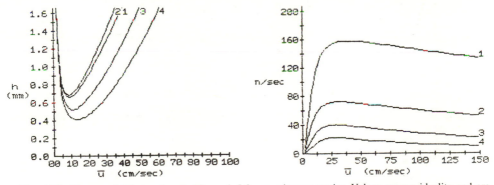

Fig. 6.53. Diameter 0.32 mm, length 30 m, d_f 5.0 μm, nitrogen carrier. Values assume ideality and are calculated for isothermal conditions. All values of h in millimeters, u_{opt} and OPGV in centimeters per second. Values for "time" are values of t_R for that solute at u_{opt} or OPGV.

| | | u_{opt} data | | | | OPGV data | | |
plot	k	h_{min}	u_{opt}	total n	time	OPGV	n/sec_{max}	total n	time
1	0.5	0.66	9	45,500	500	46	158	15,500	98
2	2	0.68	8	44,100	1125	38	72	17,100	237
3	7	0.52	11	57,700	2182	29	41	33,900	828
4	20	0.41	32	73,200	1969	37	22	37,500	1703

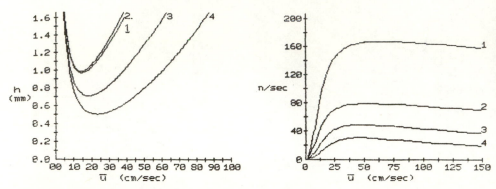

Fig. 6.54. Diameter 0.32 mm, length 30 m, d_f 5.0 μm, helium carrier. Values assume ideality and are calculated for isothermal conditions. All values of h in millimeters, u_{opt} and OPGV in centimeters per second. Values for "time" are values of t_R for that solute at u_{opt} or OPGV.

| | | u_{opt} data | | | | OPGV data | | | |
plot	k	h_{min}	u_{opt}	total n	time	OPGV	n/sec_{max}	total n	time
1	0.5	0.97	14	30,900	321	82	166	9,100	55
2	2	0.99	15	30,300	600	62	79	11,500	145
3	7	0.71	19	42,300	1263	59	49	19,900	407
4	20	0.50	23	60,000	2739	52	31	37,600	1212

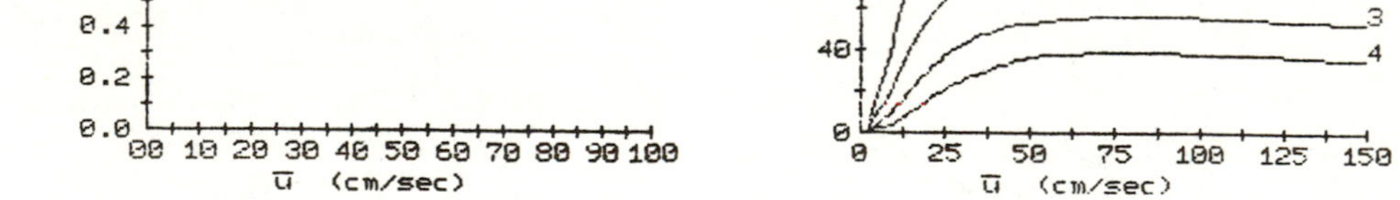

Fig. 6.55. Diameter 0.32 mm, length 30 m, d_f 5.0 μm, hydrogen carrier. Values assume ideality and are calculated for isothermal conditions. All values of h in millimeters, u_{opt} and OPGV in centimeters per second. Values for "time" are values of t_R for that solute at u_{opt} or OPGV.

| | | u_{opt} data | | | | OPGV data | | | |
plot	k	h_{min}	u_{opt}	total n	time	OPGV	n/sec_{max}	total n	time
1	0.5	1.21	18	24,800	250	148	172	5,200	30
2	2	1.22	18	24,600	500	131	83	5,700	69
3	7	0.87	25	34,500	960	99	56	13,600	242
4	20	0.59	32	50,800	1969	92	39	26,700	685

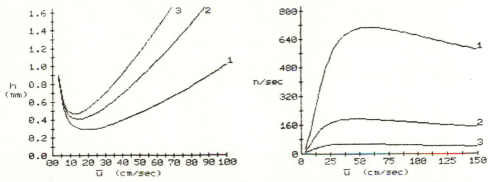

Fig. 6.56. Diameter 0.53 mm, length 15 m, d_f 1.0 μm, nitrogen carrier. Values assume ideality and are calculated for isothermal conditions. All values of h in millimeters, u_{opt} and OPGV in centimeters per second. Values for "time" are values of t_R for that solute at u_{opt} or OPGV.

| | | u_{opt} data | | | | OPGV data | | |
plot	k	h_{min}	u_{opt}	total n	time	OPGV	n/sec_{max}	total n	time
1	0.5	0.30	24	50,000	94	62	706	25,600	36
2	2	0.41	16	36,600	281	49	194	17,800	92
3	7	0.47	14	31,900	857	47	56	14,300	255

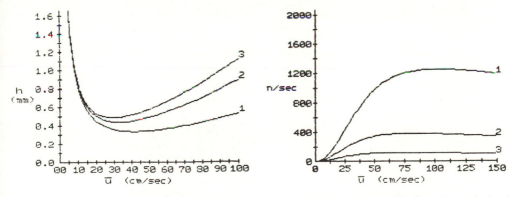

Fig. 6.57. Diameter 0.53 mm, length 15 m, d_f 1.0 μm, helium carrier. Values assume ideality and are calculated for isothermal conditions. All values of h in millimeters, u_{opt} and OPGV in centimeters per second. Values for "time" are values of t_R for that solute at u_{opt} or OPGV.

| | | u_{opt} data | | | | OPGV data | | |
plot	k	h_{min}	u_{opt}	total n	time	OPGV	n/sec_{max}	total n	time
1	0.5	0.33	43	45,500	52	103	1253	27,400	22
2	2	0.44	38	34,100	118	87	378	19,600	52
3	7	0.49	33	30,600	364	83	116	16,800	145

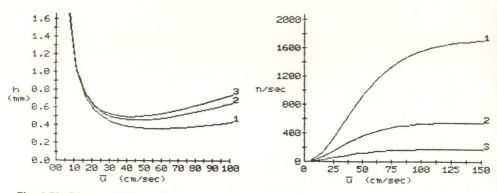

Fig. 6.58. Diameter 0.53 mm, length 15 m, d_f 1.0 μm, hydrogen carrier. Values assume ideality and are calculated for isothermal conditions. All values of h in millimeters, u_{opt} and OPGV in centimeters per second. Values for "time" are values of t_R for that solute at u_{opt} or OPGV.

| | | u_{opt} data | | | | OPGV data | | |
plot	k	h_{min}	u_{opt}	total n	time	OPGV	n/sec_{max}	total n	time
1	0.5	0.36	62	41,700	36	155	1720	10,500	15
2	2	0.46	52	32,600	87	143	538	16,900	31
3	7	0.49	42	30,600	287	128	173	16,200	94

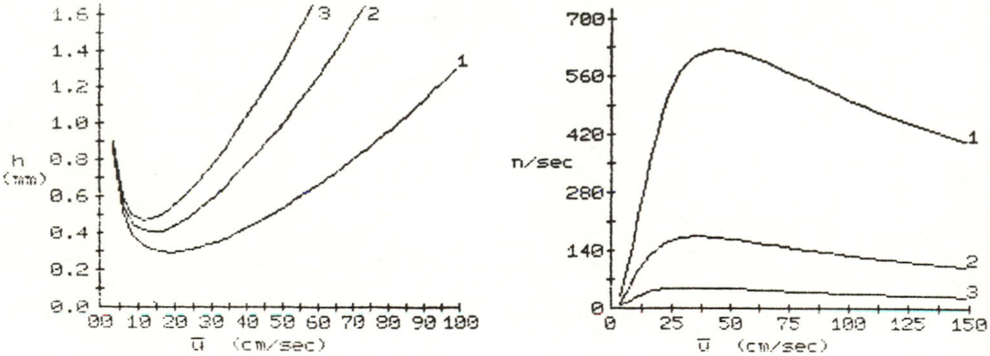

Fig. 6.59. Diameter 0.53 mm, length 30 m, d_f 1.0 μm, nitrogen carrier. Values assume ideality and are calculated for isothermal conditions. All values of h in millimeters, u_{opt} and OPGV in centimeters per second. Values for "time" are values of t_R for that solute at u_{opt} or OPGV.

| | | u_{opt} data | | | | OPGV data | | |
plot	k	h_{min}	u_{opt}	total n	time	OPGV	n/sec_{max}	total n	time
1	0.5	0.29	19	103,000	237	44	625	63,900	102
2	2	0.41	16	73,000	563	39	173	39,900	231
3	7	0.47	11	63,800	2182	39	50	30,800	615

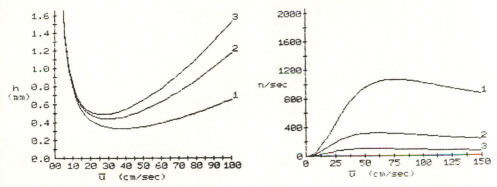

Fig. 6.60. Diameter 0.53 mm, length 30 m, d_f 1.0 μm, helium carrier. Values assume ideality and are calculated for isothermal conditions. All values of h in millimeters, u_{opt} and OPGV in centimeters per second. Values for "time" are values of t_R for that solute at u_{opt} or OPGV.

		u_{opt} data				OPGV data		
plot	k	h_{min}	u_{opt}	total n	time	OPGV	n/sec$_{max}$	total n	time
1	0.5	0.33	42	90,900	107	76	1064	63,000	59
2	2	0.44	33	68,200	273	61	322	47,500	148
3	7	0.49	31	61,200	774	61	98	38,600	393

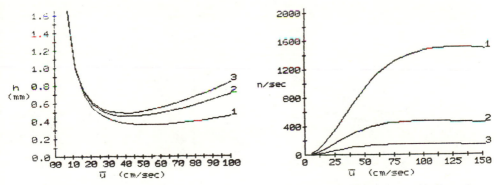

Fig. 6.61. Diameter 0.53 mm, length 30 m, d_f 1.0 μm, hydrogen carrier. Values assume ideality and are calculated for isothermal conditions. All values of h in millimeters, u_{opt} and OPGV in centimeters per second. Values for "time" are values of t_R for that solute at u_{opt} or OPGV.

		u_{opt} data				OPGV data		
plot	k	h_{min}	u_{opt}	total n	time	OPGV	n/sec$_{max}$	total n	time
1	0.5	0.36	62	83,300	73	130	1510	52,300	35
2	2	0.46	47	65,200	191	111	474	38,400	81
3	7	0.49	42	61,200	571	96	151	37,800	250

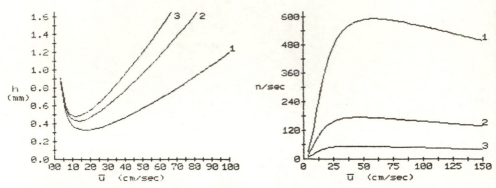

Fig. 6.62. Diameter 0.53 mm, length 15 m, d_f 1.5 μm, nitrogen carrier. Values assume ideality and are calculated for isothermal conditions. All values of h in millimeters, u_{opt} and OPGV in centimeters per second. Values for "time" are values of t_R for that solute at u_{opt} or OPGV.

			u_{opt} data				OPGV data		
plot	k	h_{min}	u_{opt}	total n	time	OPGV	n/sec_{max}	total n	time
1	0.5	0.33	19	45,500	118	62	591	21,400	36
2	2	0.43	14	34,900	321	52	175	15,100	87
3	7	0.48	11	31,300	1091	44	54	14,700	273

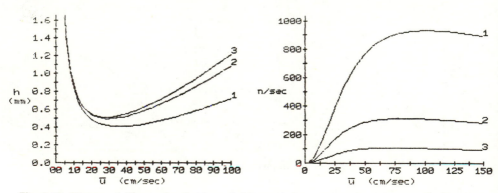

Fig. 6.63. Diameter 0.53 mm, length 15 m, d_f 1.5 μm, helium carrier. Values assume ideality and are calculated for isothermal conditions. All values of h in millimeters, u_{opt} and OPGV in centimeters per second. Values for "time" are values of t_R for that solute at u_{opt} or OPGV.

			u_{opt} data				OPGV data		
plot	k	h_{min}	u_{opt}	total n	time	OPGV	n/sec_{max}	total n	time
1	0.5	0.41	43	36,600	52	103	930	20,300	22
2	2	0.49	29	30,600	155	87	313	16,200	52
3	7	0.51	31	29,400	387	83	107	15,500	144

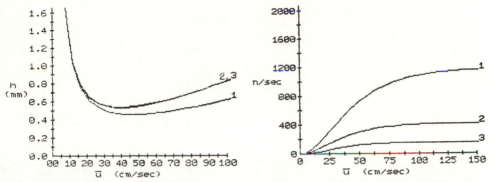

Fig. 6.64. Diameter 0.53 mm, length 15 m, d_f 1.5 μm, hydrogen carrier. Values assume ideality and are calculated for isothermal conditions. All values of h in millimeters, u_{opt} and OPGV in centimeters per second. Values for "time" are values of t_R for that solute at u_{opt} or OPGV.

| | | u_{opt} data | | | | OPGV data | | |
plot	k	h_{min}	u_{opt}	total n	time	OPGV	n/sec$_{max}$	total n	time
1	0.5	0.46	52	32,600	43	155	1130	16,400	15
2	2	0.54	42	27,800	107	143	415	13,100	31
3	7	0.53	42	28,300	286	128	154	14,400	94

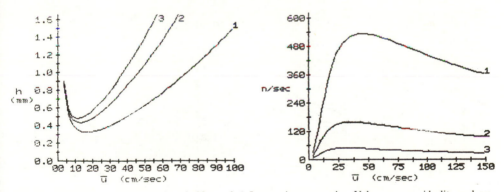

Fig. 6.65. Diameter 0.53 mm, length 30 m, d_f 1.5 μm, nitrogen carrier. Values assume ideality and are calculated for isothermal conditions. All values of h in millimeters, u_{opt} and OPGV in centimeters per second. Values for "time" are values of t_R for that solute at u_{opt} or OPGV.

| | | u_{opt} data | | | | OPGV data | | |
plot	k	h_{min}	u_{opt}	total n	time	OPGV	n/sec$_{max}$	total n	time
1	0.5	0.33	19	90,900	237	46	532	52,000	98
2	2	0.43	14	69,800	643	39	158	36,500	231
3	7	0.48	11	62,500	2182	41	48	28,100	585

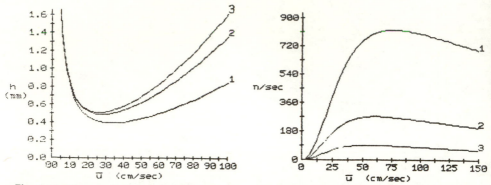

Fig. 6.66. Diameter 0.53 mm, length 30 m, d_f 1.5 μm, helium carrier. Values assume ideality and are calculated for isothermal conditions. All values of h in millimeters, u_{opt} and OPGV in centimeters per second. Values for "time" are values of t_R for that solute at u_{opt} or OPGV.

		u_{opt} data				OPGV data			
plot	k	h_{min}	u_{opt}	total n	time	OPGV	n/sec_{max}	total n	time
1	0.5	0.40	40	75,000	113	76	822	48,700	59
2	2	0.49	31	61,200	290	64	273	38,400	141
3	7	0.51	29	58,800	828	59	92	37,400	407

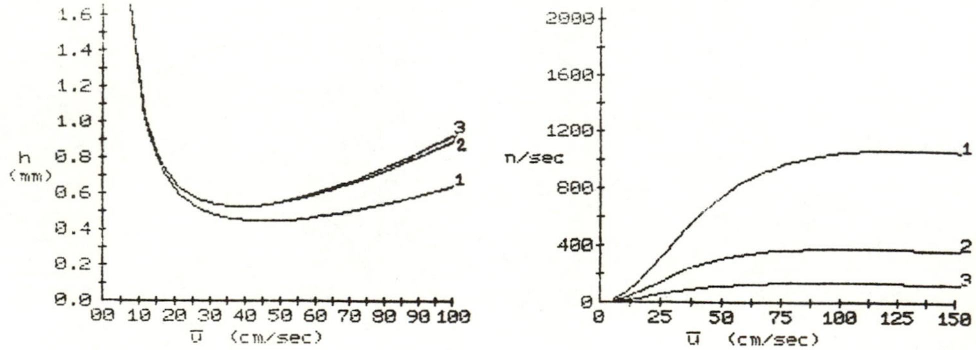

Fig. 6.67. Diameter 0.53 mm, length 30 m, d_f 1.5 μm, hydrogen carrier. Values assume ideality and are calculated for isothermal conditions. All values of h in millimeters, u_{opt} and OPGV in centimeters per second. Values for "time" are values of t_R for that solute at u_{opt} or OPGV.

		u_{opt} data				OPGV data			
plot	k	h_{min}	u_{opt}	total n	time	OPGV	n/sec_{max}	total n	time
1	0.5	0.45	52	66,700	87	130	1065	36,900	35
2	2	0.53	42	56,600	214	106	376	31,900	85
3	7	0.53	42	56,600	571	101	136	32,300	238

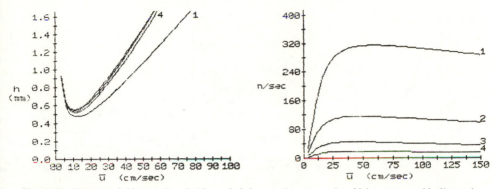

Fig. 6.68. Diameter 0.53 mm, length 15 m, d_f 3.0 μm, nitrogen carrier. Values assume ideality and are calculated for isothermal conditions. All values of h in millimeters, u_{opt} and OPGV in centimeters per second. Values for "time" are values of t_R for that solute at u_{opt} or OPGV.

| | | u_{opt} data | | | | OPGV data | | |
plot	k	h_{min}	u_{opt}	total n	time	OPGV	n/sec$_{max}$	total n	time
1	0.5	0.48	14	31,300	161	64	314	11,000	35
2	2	0.55	11	27,300	409	57	115	9,100	79
3	7	0.54	11	27,800	1091	44	45	12,300	273
4	20	0.52	11	28,900	2864	67	18	8,400	470

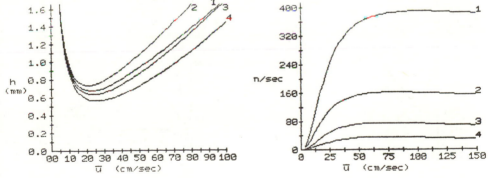

Fig. 6.69. Diameter 0.53 mm, length 15 m, d_f 3.0 μm, helium carrier. Values assume ideality and are calculated for isothermal conditions. All values of h in millimeters, u_{opt} and OPGV in centimeters per second. Values for "time" are values of t_R for that solute at u_{opt} or OPGV.

| | | u_{opt} data | | | | OPGV data | | |
plot	k	h_{min}	u_{opt}	total n	time	OPGV	n/sec$_{max}$	total n	time
1	0.5	0.68	24	22,100	94	111	389	7,900	20
2	2	0.74	22	20,300	205	94	162	7,800	48
3	7	0.64	26	23,400	462	85	76	10,700	141
4	20	0.57	29	26,300	1026	87	35	12,700	362

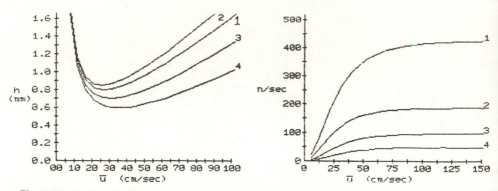

Fig. 6.70. Diameter 0.53 mm, length 15 m, d_f 3.0 μm, hydrogen carrier. Values assume ideality and are calculated for isothermal conditions. All values of h in millimeters, u_{opt} and OPGV in centimeters per second. Values for "time" are values of t_R for that solute at u_{opt} or OPGV.

		u_{opt} data				OPGV data			
plot	k	h_{min}	u_{opt}	total n	time	OPGV	n/sec_{max}	total n	time
1	0.5	0.80	27	18,800	83	158	425	6,100	14
2	2	0.85	27	17,600	167	147	186	5,700	31
3	7	0.70	32	21,400	375	85	76	10,700	141
4	20	0.60	42	25,000	750	118	49	13,100	267

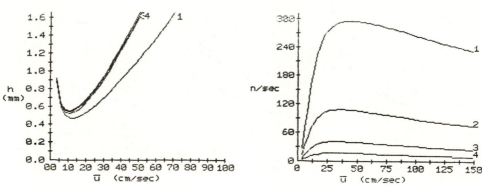

Fig. 6.71. Diameter 0.53 mm, length 30 m, d_f 3.0 μm, nitrogen carrier. Values assume ideality and are calculated for isothermal conditions. All values of h in millimeters, u_{opt} and OPGV in centimeters per second. Values for "time" are values of t_R for that solute at u_{opt} or OPGV.

		u_{opt} data				OPGV data			
plot	k	h_{min}	u_{opt}	total n	time	OPGV	n/sec_{max}	total n	time
1	0.5	0.47	14	63,800	321	46	297	29,100	98
2	2	0.55	11	54,600	818	39	108	24,900	231
3	7	0.54	11	55,600	2182	39	41	25,200	615
4	20	0.52	11	57,700	5727	39	17	27,500	1615

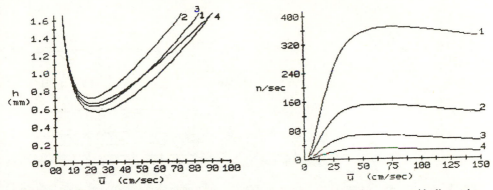

Fig. 6.72. Diameter 0.53 mm, length 30 m, d_f 3.0 μm, helium carrier. Values assume ideality and are calculated for isothermal conditions. All values of h in millimeters, u_{opt} and OPGV in centimeters per second. Values for "time" are values of t_R for that solute at u_{opt} or OPGV.

		u_{opt} data				OPGV data		
plot	k	h_{min}	u_{opt}	total n	time	OPGV	n/sec_{max}	total n	time
1	0.5	0.66	24	45,400	188	80	369	20,800	56
2	2	0.72	22	41,700	409	66	151	20,600	136
3	7	0.63	24	47,600	1000	61	68	24,000	393
4	20	0.56	24	53,600	2625	55	31	35,500	1145

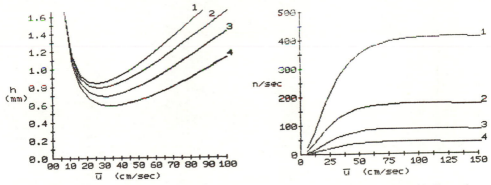

Fig. 6.73. Diameter 0.53 mm, length 30 m, d_f 3.0 μm, hydrogen carrier. Values assume ideality and are calculated for isothermal conditions. All values of h in millimeters, u_{opt} and OPGV in centimeters per second. Values for "time" are values of t_R for that solute at u_{opt} or OPGV.

		u_{opt} data				OPGV data		
plot	k	h_{min}	u_{opt}	total n	time	OPGV	n/sec_{max}	total n	time
1	0.5	0.79	27	38,000	167	139	411	13,300	32
2	2	0.84	27	35,700	333	125	177	12,700	72
3	7	0.69	32	43,500	750	111	89	19,200	216
4	20	0.59	37	50,800	1703	96	44	28,900	656

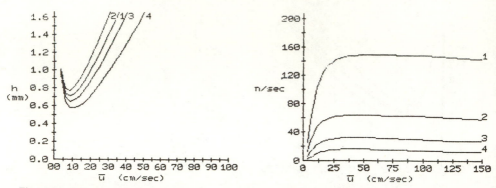

Fig. 6.74. Diameter 0.53 mm, length 15 m, d_f 5.0 μm, nitrogen carrier. Values assume ideality and are calculated for isothermal conditions. All values of h in millimeters, u_{opt} and OPGV in centimeters per second. Values for "time" are values of t_R for that solute at u_{opt} or OPGV.

		u_{opt} data			OPGV data			
plot	k	h_{min}	u_{opt}	total n	time	OPGV	n/sec_{max}	total n	time
1	0.5	0.71	9	21,100	250	76	149	4,400	30
2	2	0.77	9	19,500	500	64	64	4,500	70
3	7	0.65	9	23,100	1333	59	32	6,500	203
4	20	0.58	11	25,900	2864	54	16	9,300	583

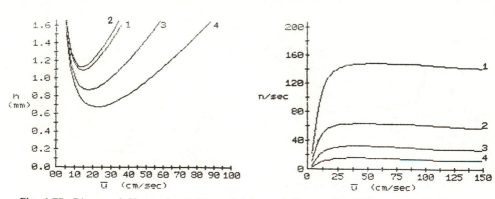

Fig. 6.75. Diameter 0.53 mm, length 15 m, d_f 5.0 μm, helium carrier. Values assume ideality and are calculated for isothermal conditions. All values of h in millimeters, u_{opt} and OPGV in centimeters per second. Values for "time" are values of t_R for that solute at u_{opt} or OPGV.

		u_{opt} data			OPGV data			
plot	k	h_{min}	u_{opt}	total n	time	OPGV	n/sec_{max}	total n	time
1	0.5	1.09	15	13,800	150	65	149	5,200	35
2	2	1.13	15	13,300	300	56	64	5,100	80
3	7	0.87	17	17,200	705	52	32	7,400	231
4	20	0.68	24	22,100	1313	47	16	10,700	670

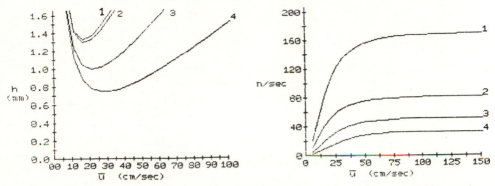

Fig. 6.76. Diameter 0.53 mm, length 15 m, d_f 5.0 μm, hydrogen carrier. Values assume ideality and are calculated for isothermal conditions. All values of h in millimeters, u_{opt} and OPGV in centimeters per second. Values for "time" are values of t_R for that solute at u_{opt} or OPGV.

| | | u_{opt} data | | | | OPGV data | | |
plot	k	h_{min}	u_{opt}	total n	time	OPGV	n/sec$_{max}$	total n	time
1	0.5	1.30	16	11,500	141	160	168	2,400	14
2	2	1.33	16	11,300	281	150	80	3,800	30
3	7	1.00	22	15,000	545	150	48	3,800	80
4	20	0.75	32	20,000	984	145	31	6,700	217

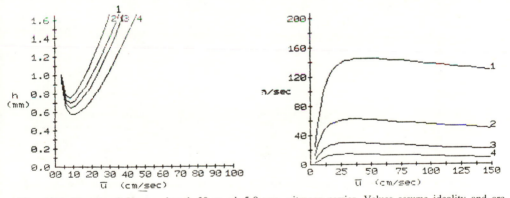

Fig. 6.77. Diameter 0.53 mm, length 30 m, d_f 5.0 μm, nitrogen carrier. Values assume ideality and are calculated for isothermal conditions. All values of h in millimeters, u_{opt} and OPGV in centimeters per second. Values for "time" are values of t_R for that solute at u_{opt} or OPGV.

| | | u_{opt} data | | | | OPGV data | | |
plot	k	h_{min}	u_{opt}	total n	time	OPGV	n/sec$_{max}$	total n	time
1	0.5	0.70	9	42,900	500	56	145	11,700	80
2	2	0.76	9	39,500	1000	41	62	13,600	220
3	7	0.65	9	46,200	2667	44	30	16,400	545
4	20	0.58	11	51,700	5727	34	15	27,800	1853

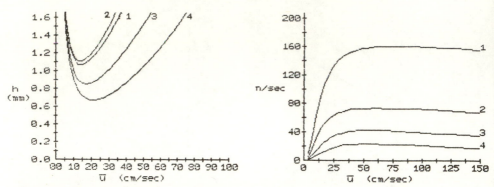

Fig. 6.78. Diameter 0.53 mm, length 30 m, d_f 5.0 μm, helium carrier. Values assume ideality and are calculated for isothermal conditions. All values of h in millimeters, u_{opt} and OPGV in centimeters per second. Values for "time" are values of t_R for that solute at u_{opt} or OPGV.

| | | u_{opt} data | | | | OPGV data | | |
plot	k	h_{min}	u_{opt}	total n	time	OPGV	n/sec$_{max}$	total n	time
1	0.5	1.07	15	28,000	300	97	160	7,400	46
2	2	1.11	15	27,000	600	64	74	10,400	141
3	7	0.85	17	35,300	1411	72	42	14,000	333
4	20	0.67	22	44,800	2864	62	25	25,400	1016

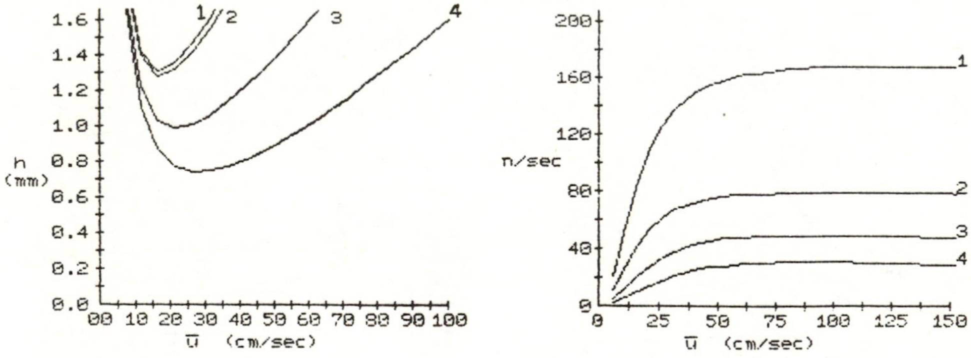

Fig. 6.79. Diameter 0.53 mm, length 30 m, d_f 5.0 μm, hydrogen carrier. Values assume ideality and are calculated for isothermal conditions. All values of h in millimeters, u_{opt} and OPGV in centimeters per second. Values for "time" are values of t_R for that solute at u_{opt} or OPGV.

| | | u_{opt} data | | | | OPGV data | | |
plot	k	h_{min}	u_{opt}	total n	time	OPGV	n/sec$_{max}$	total n	time
1	0.5	1.28	16	23,400	281	155	160	4,600	29
2	2	1.31	16	22,900	562	148	79	4,800	61
3	7	0.99	21	30,300	1143	134	49	8,800	179
4	20	0.74	27	40,500	2333	106	31	18,400	594

column available, (3) the stationary phase film thickness, (4) the type of stationary phase, (5) the carrier gas employed, and (6) the partition ratios of the solutes
for which the best separation is desired.

Values of $\bar{u}_{opt}$ should be lowered slightly for columns with thicker films of
stationary phase and/or for columns with films thicker than ~0.5 μm of stationary phases that contain appreciable amounts of cyanopropyl (D_S is lower).

The graphs presented for microbore columns are restricted to hydrogen carrier
gas; helium is an impractical carrier for these very fine diameter columns.
Graphs for the longer microbore columns will be of very limited utility in most
cases because of the very high head pressures that would be required; even a 10-
m microbore column may demand head pressures of 60–70 psi hydrogen.

Until d_f exceeds ~2 μm, plots for $k = 7$ and higher values of k are very
similar. Hence most of the following graphs illustrate three plots, representing
low ($k = 0.5$), intermediate ($k = 2$), and high ($k = 7$) solute partition ratios. The
increase in h_{min} with increasing d_f is much more striking for lower-k solutes; in
addition, solute partition ratios increase on thicker-film columns if other parameters are held constant. Graphs for the 3- and 5-μm film columns therefore
include a fourth plot at $k = 20$. Although this condition may prevail during most
of the chromatographic process, the column temperature would normally be
increased to elute that solute at a lower k to enhance sensitivity.

6.10   Columns for Mass Spectrometry

The combination of gas chromatography with mass spectrometry is a powerful
one. If compounds are to be characterized on the basis of their mass spectral
fragmentation patterns, results are most useful if the spectrum is limited to ions
derived from a single molecular species; gas chromatography, as the most
powerful separation technique, has a high potential for supplying the mass spectrometer with one molecular species at a time. In addition, the two modes of
characterization (gas chromatographic retentions and mass spectral patterns) are
truly orthogonal (i.e., they depend on entirely different molecular properties);
this greatly increases the reliability of identifications that are based on both
retention characteristics in a high-resolution system and matching fragmentation
patterns.

In most applications, the mass spectrometer can be envisioned as a specialized
detector, attached to the end of the chromatographic column. This detector is
intolerant of background signal, and even moderate column bleed levels may
require excessive signal attenuation. Also, it operates under high-vacuum conditions, and the gas flow that can be accepted is governed largely by the pumping
capacity. The column may be connected through some type of separator or open
split that diverts the major portion of the carrier gas, or it can be plumbed directly
into the ion source. The latter alternative requires sufficient pumping capacity on

the part of the mass spectrometer to maintain the necessary vacuum while accepting the total carrier flow through the column; this, in turn, requires that the pressure drop through the column be sufficient to ensure both good chromatography in the column and adequate vacuum in the mass spectrometer.

The pressure drop through a 530-μm-diameter column is usually too low to permit it to be connected directly into the ion source of a conventional mass spectrometer; as is the case with packed columns, some type of pressure restriction or interface will be required. In general, the pressure drop required for direct interfacing can be provided by least 20 and preferably 30 m of 320-μm column, or 15 and preferably 25–30 m of 0.25-μm column.

Direct connection of the column into the high-vacuum region of the mass spectrometer also requires vacuum-tight connections. The porosity of standard graphite ferrules is slight and they are rarely faulted for general chromatographic purposes, where their superior compressibility is used to advantage. That porosity is sufficient to cause problems under high-vacuum conditions, and most authorities prefer graphitized Vespel ferrules for the high-vacuum column connection. The limited compressibility of these ferrules demands more precise tolerances between the ferrule hole and the column outer diameter and may require sizing with jeweler's drills.

References

1. K. Yabumoto, D. K. Ingraham, and W. Jennings, *J. High Res. Chromatogr.* **3,** 248 (1980).
2. W. Jennings, "Gas Chromatography with Glass Capillary Columns," 2nd ed. Academic Press, New York, 1980.
3. W. Jennings and T. Shibamoto, "Qualitative Analysis of Flavor and Fragrance Volatiles by Glass Capillary Gas Chromatography." Academic Press, New York, 1980.
4. M. F. Mehran, W. J. Cooper, R. Lautamo, R. R. Freemen, and W. Jennings, *J. High Res. Chromatogr.* **8,** 715 (1985).
5. D. F. Ingraham, C. F. Shoemaker, and W. Jennings, *J. Chromatogr.* **239,** 39 (1982).
6. G. Takeoka, H. M. Richard, M. F. Mehran, and W. Jennings, *J. High Res. Chromatogr.* **6,** 145 (1983).
7. M. F. Mehran, W. J. Cooper, and W. Jennings, *J. High Res. Chromatogr.* **7,** 215 (1984).
8. G. Takeoka and W. Jennings, *J. Chromatogr. Sci.* **22,** 177 (1984).
9. M. F. Mehran, W. J. Cooper, M. Mehran, and W. Jennings, *J. Chromatogr. Sci.* **24,** 142 (1986).
10. W. Jennings and G. Takeoka, *in* "Analysis of Volatiles: Methods and Applications" (P. Schreier, ed.), p. 63. de Gruyter, Berlin, 1984.
11. W. Jennings, *Am. Lab.* **16,** 14 (1984).
12. A. E. Coleman, *J. Chromatogr. Sci.* **13,** 198 (1973).
13. K. Grob and G. Grob, *J. High Res. Chromatogr.* **5,** 349 (1982).
14. G. Schomburg, R. Dielmann, H. Borowski, and H. Husmann, *J. Chromatogr.* **167,** 337 (1978).
15. V. Paramasigamani and W. Aue, *J. Chromatogr.* **168,** 202 (1979).
16. W. Jennings, *J. Chromatogr. Sci.* **21,** 337 (1983).

CHAPTER 7

INSTRUMENT CONVERSION AND ADAPTATION

7.1 General Considerations

A large percentage of all the gas chromatographs that are still in use today were originally supplied as packed column instruments; some of these have been converted for operation with open tubular columns, but most are still dedicated to packed column use. In the majority of cases, conversion of the packed column instrument to open tubular columns has become a remarkably simple exercise. In years past, such conversions necessitated the purchase of optional capillary accessories from the instrument manufacturer and/or considerable ingenuity on the part of the researcher. A number of the kits that have recently become available greatly simplify retrofitting the packed column instrument. The performance of even an older instrument will be much improved by conversion to open tubular columns, provided the response of the electronics is sufficiently fast [1].

7.2 Oven Considerations

The fused silica open tubular column is characterized by a very low thermal mass and responds very quickly when exposed to radiant heat. The heating coils of most modern instruments are mounted behind a barrier that prevents radiant heat from reaching the column. In some other instruments, a line of sight exists between the column and the oven heater, and some degree of difficulty can be experienced. Especially when the column is mounted under conditions where its exposure to the heat source is nonuniform, chromatographing bands are repetitively exposed to relatively hot and relatively cool sections of the column in an alternating sequence. If the front of a chromatographing band is exposed to a higher temperature and the rear of that band is decelerated by a lower temperature, split or malformed peaks (e.g., the "Christmas tree" effect [2,3]) can result. A simple barrier of aluminum foil has been used to block the heater-to-column line of sight and rectify this problem by shielding the column from radiant heat. The barrier should be placed where it does not interfere with the flow of oven air; some prefer to loosely shroud the column with foil [4]. Foil barriers are neither necessary nor desirable in other cases.

Some attention should also be directed to suspension of the column within the oven. Column-mounting hardware should support the column cage and preferably should not come into contact with the column; neither should any portion of the column be allowed to come in contact with oven walls. With the exception of those portions attached to the inlet and to the detector, the low thermal mass column should be able to follow the temperature of the oven air without restriction. It is also important to secure the column-mounting hardware in a position that permits unstrained connections to the injector and detector (see Chapter 6).

7.3   Carrier Gas Considerations

Many of the advantages offered by modern open tubular columns are attributable to improved methods of tubing deactivation and to the high-purity cross-linked, surface-bonded polymers with which the tubing is coated. On the other hand, because their tolerance for carrier gas-borne impurities is much lower, their chromatographic performance and longevity are more strongly influenced by the carrier gas quality. To ensure optimum performance and extended column life, instrument conversion should also involve a critical evaluation of the gas lines and gas purification devices.

Because of these considerations, the premium grades of carrier gas are usually less expensive in the long run. Stainless steel diaphragm regulators and flow controllers are preferred; oxygen contamination by diffusion through neoprene diaphragms and O-rings has been documented many times.

The superiority of hydrogen as a carrier was discussed in Chapter 5; the safety of a hydrogen-supplied installation can be improved by employing a separate mass flow controller for each hydrogen demand function. Ideally, a two-stage stainless steel diaphragm pressure regulator is installed on the cylinder, followed by both a hydrocarbon trap and a high-capacity oxygen scrubber. Stainless steel diaphragm mass flow controllers connected in parallel (rather than in series) can now be used to direct the reduced-pressure hydrogen to each function. That for a flame ionization detector is set to deliver ~ 30 cm^3/min; that supplying an on-column injector attached to a 0.32-mm capillary may be set to 3 cm³/min. Even in the event of a broken column or a ruptured line, hydrogen discharge is limited to the setting of that particular mass flow controller.

Carrier gas lines should be metal rather than plastic; one model of retractable on-column injector requires some flexibility on the carrier gas line and is normally supplied with a length of heavy-walled Teflon. On GC/MS installations where that has been replaced with a length of loosely coiled ⅟₁₆-in. stainless steel tubing, a marked reduction in oxygen has been noted.

The degree to which oxygen causes column deterioration is influenced by the type of stationary phase and the temperatures to which the column is exposed. These interrelationships are discussed further in Chapter 10; for systems that are susceptible, some insurance against oxygen entry through leaky fittings is desirable. Additional indicating oxygen scrubbers should also be installed immediately before each inlet system to minimize the amount of oxygen coming in contact with the column. Because many parts of the system (including septa, flow controller O-ring seals, pressure regulator diaphragms, elastomeric lines) are permeable to oxygen, a truly oxygen-free system is rarely if ever possible.

7.4 Packed to Large-Diameter Open Tubular Conversion

The simplest conversion is invariably from the packed column to the 530-μm large-diameter open tubular column, because that column can utilize the same conventional packed column injector and can also be used without detector modification. The much higher gas flow volumes that are often used with these large-diameter columns duplicate those used by the packed column and impose no additional demands in terms of operator technique or training. The benefits of this conversion include not only the instrumental simplicity and user-friendliness cited above but also the ruggedness of these columns, which holds strong appeal for those with long memories for the fragility of the conventional glass capillary column.

The larger-diameter column, of course, encompasses a larger volume of mobile (gas) phase per unit of column length, and to maintain a more nearly normal phase ratio, thicker films of stationary phase are required; this thicker film further benefits the inertness of the columns. In addition, the columns are normally used with conventional packed column injectors and operated without splitting or purging; i.e., the entire sample is passed through the column to the detector. Both of these factors—the greater inertness of the column and direct undiverted injection—increase the quantitative reliability of the system.

Oven Size Considerations

The first requirement for any conversion, of course, is that the column can be physically fitted into the oven of the chromatograph; this was formerly a problem with the large-diameter fused silica columns. Current procedures require drawing the fused silica as straight tubing, which is then forced to a coiled configuration. Released from all constraints, the tubing straightens, which facilitates proper connections of the column to the inlet and detector but also serves to remind the analyst that the coiled column is in a stressed configuration. The diameter to which these large-diameter columns are coiled varies from supplier to supplier. As explored in Chapter 2, the rate of growth of the surface flaws in the tubing is proportional to the degree of tubing strain. The strain on the tubing increases as the coiling diameter decreases; only the highest-quality tubing (in terms of surface flaws, internal strains, and polyimide coating) can be coiled to the smaller diameters. Some large-diameter columns are available only with a coil diameter of approximately 30 cm (12 in.) and can be installed only in chromatographs with exceptionally large ovens. Other such columns have a coil diameter of 18 cm (8 in.) and can be used in all but two or three models of very small-oven chromatographs. Owners of the latter are sometimes tempted to rewind commercial columns to smaller coil diameters, and some success has been achieved. The better grades of 530-μm tubing available at this writing can be coiled even more tightly: test coils of 7.5 cm (3 in.) have endured for extended periods at room temperature [5], but the additional stress of higher temperature would almost surely lead to frequent breakage.

Inlet Conversion

Conversion kits are available from several supply houses. The simpler kits provide a glass injector liner and adapters that reduce the column connections on the packed column injector and detector to $\frac{1}{16}$-in. fittings. Even the simplest conversion kits are usually satisfactory if these columns are limited to high-flow, low-resolution mode operation and if the size of the injected sample is restricted to about 1 μl. Proper adapter design, which is discussed in Chapter 3, becomes increasingly important with larger sample injections and/or higher resolution (low-flow) operation. Figure 7.1 shows an adapter kit designed for the HP 5880, and one suitable for bottom-sealing packed column injectors terminating in a $\frac{1}{4}$-in. Swagelok fitting (e.g., HP 5700, Varian 3700) is shown in Fig. 7.2. For the analysis of sensitive compounds, it can be advantageous to deactivate the sample-contacting glass liner; a suitable deactivation procedure is described in Section 7.7.

Detector Conversion

Most adapters for the detector side are similar to a reducing tee, whose side-arm is normally capped when the column is operated in high-flow mode. With FID, the outlet end of the column should terminate inside the flame jet so it is swept by the combustion hydrogen (Fig. 7.3 [6]). Some flame jets have internal diameters that are too small to accept large-diameter open tubular columns; these require a "butt-type" junction, which is usually suitable in high-flow mode but

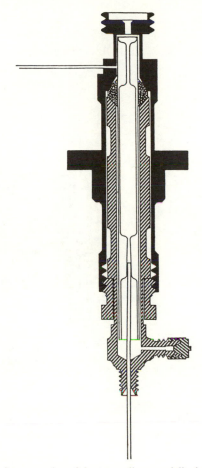

Fig. 7.1. Inlet adapters for conversion of the top-sealing, specially threaded HP 5880 injector to large-diameter open tubular column capability. (Drawing by R. L. Masterson. Reprinted with permission of the copyright holder, J&W Scientific, Inc.)

which may lead to peak broadening in low-flow mode, and active sites within the flame jet may cause problems. Several jet options are available for some popular instruments and can be provided by the instrument supplier on inquiry.

Makeup Gas Considerations

The large-diameter open tubular column can be operated without makeup gas by capping the adapter sidearm (Fig. 7.3); alternatively, that port can be used for the addition of makeup gas. Makeup gas can serve two useful purposes:

1. In cases where solutes are eluted under conditions of low volumetric flow, closely eluting bands have opportunities to remix before they reach the zone of

Fig. 7.2. Inlet adapter typical of those used with the HP 5700, Varian 3700, and other instruments with ¼-in. Swagelok fittings. (Drawing by R. Materson. Reproduced with permission of the copyright holder, J&W Scientific, Inc.)

detection, and they also stay too long in the detector; this results in broadened peaks and loss of resolution. These problems are minimized in a well-designed FID by housing the end of the column within the flame jet. Eluting solutes are discharged into the stream of combustion hydrogen, ensuring their rapid transport to and from the zone of detection and producing clean, sharp peaks from short chromatographic bands.

2. Makeup gas may permit the detector to be operated in a plateau region of

Fig. 7.3. Typical detector adapter, suitable for conversion of the packed column instrument to either large-diameter (0.53 mm) or standard capillary open tubular columns [6]. The column should normally be housed within the flame jet. The sidearm provides the entry point for makeup gas (when used) and can be capped for large-diameter columns in high-flow mode.

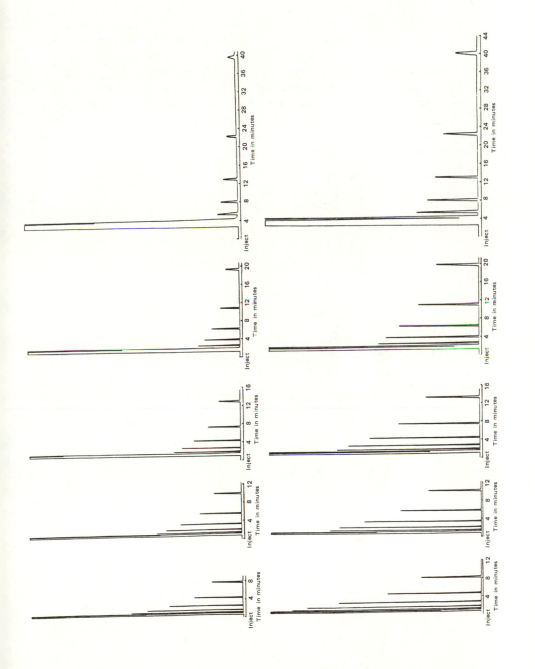

high sensitivity [7]. Most detectors are designed to operate under packed column flow conditions, and sensitivity can be seriously eroded by lower carrier gas flows or by changes in the ratios of the detector gases (for FID, this would be the ratio carrier : combustion hydrogen : air). Most FIDs exhibit reasonably flat response (i.e., a plateau) and optimum sensitivity at gas flow ratios approximating 1 : 1 : 10 for nitrogen, hydrogen, and air, respectively. In addition, the total flow volume must be commensurate with the size of the jet orifice: the linear flow velocity through the orifice must be sufficient to maintain the flame, but with a given orifice size, higher flow volumes result in linear velocities so high that the flame iw self-extinguishing, i.e., it blows out.

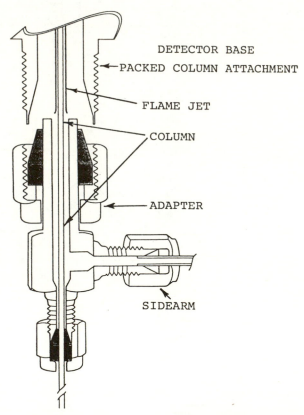

Fig. 7.4. Effect of makeup gas on the performance of large-diameter open tubular columns [6]. Column, 30 m × 0.53 mm, with helium flow at (left to right) 10, 8, 6, 4, and 2 ml/min; top chromatograms without makeup gas. Nitrogen makeup at 30 ml/min was added for the bottom chromatograms. See text for discussion. (Drawing by R. L. Masterson. Reproduced with permission of the copyright holder, J&W Scientific, Inc.)

The first makeup gas benefit described above is rarely applicable to large-diameter open tubular columns, but advantages are derived from the second. As shown in Fig. 7.4 [6], makeup gas has little or no effect on peak shape, even in low-flow mode, provided the outlet end of the column is housed within the flame jet. It can, however, produce enhanced sensitivity. Makeup gas is discussed in greater detail in a later section.

7.5 Packed to Capillary Conversion

Many aspects of this subject have been considered previously [1]; in general, the conversion will require the addition of a capillary-compatible inlet and addition of makeup gas at the detector. The latter point was discussed above, and the various types of capillary-compatible inlets are compared in Chapter 3. Kits that permit low-cost retrofitting of split, splitless, and on-column inlets to almost any packed column instrument are available from several supply houses.

7.6 Makeup Gas Considerations

Because the C_M term of the van Deemter equation is unimportant for the packed column, nitrogen is often used as the mobile phase in packed column gas chromatography; the FID gases under these conditions are (approximately) 30 cm^3/min nitrogen carrier, 30 cm^3/min combustion hydrogen, and 300 cm^3/min air. In the open tubular column, it is the C_M term that is limiting and hydrogen and helium become the superior carriers. By using hydrogen or helium as carrier while employing nitrogen as a makeup gas, detector sensitivity can be maintained while enjoying superior open tubular gas chromatography. The type and amount of makeup gas should be selected with a view to the recommendations of the detector manufacturer (use as makeup what they recommend as packed column carrier) and Table 7.1.

TABLE 7.1

Makeup Gas Recommendations

Detector	Makeup gas	Total flow, makeup + column (cm^3/min)
Flame ionization	Nitrogen	30–40
Thermal conductivity	Same as carrier	<5[a]
Electron capture	Nitrogen/argon + methane	30–60[a]
Thermionic	Helium	10–30
Flame photometric	Helium	30–60
Hall	Helium	20–40

[a]Determined by detector volume; consult detector manual.

7.7   Inlet Deactivation

The column is usually blamed for problems which are sometimes evidenced by active solutes, including solute tailing (usually attributable to a reversible adsorption) and partial or total loss of the solute (irreversible adsorption). Where this is, in fact, the fault of the column, corrective measures can sometimes be taken; these are considered in Chapter 10; frequently, however, the problem does not lie in the column but is associated with active sites in the inlet.

The vaporizing forms of injection include direct flash vaporization and hot on-column injection with the large-diameter (e.g., Megabore) open tubular columns and split, splitless, and programmed temperature vaporization with smaller-diameter (capillary) open tubular columns. The vaporization chamber is generally glass and is removable to facilitate the removal of injection residues. Materials allowed to remain in the heated inlet continue to oxidize and generally become acidic; not only do these residues react with solutes of later injections, but their continued degradation can generate baseline problems and eventually "poison" the column (see Chapter 10). Inlets are probably best cleaned by immersion in concentrated HNO_3 and rinsed in deionized water. Dichromate cleaning solutions should be avoided, as the high concentration of chromium ion may complicate later steps.

Even a clean inlet can cause activity problems; conventional glass contains both appreciable amounts of metal oxides and an abundance of surface silanols (Chapter 2); those constructed of quartz contain less metal ion, but it is sufficient to cause problems. Left untreated, inlets from either material can generate chromatograms testifying to excessive activity, even with the most inert fused silica column. The body of knowledge accumulated from problems encountered (and eventually solved) with conventional glass capillary columns can now stand us in good stead. Metal ions in the glass surface can be removed by a leaching process; glass wool and glass beads that may be used in the linear should also be treated and can be deactivated separately or (if they do not restrict reagent access) preinstalled in the liner. Of course, great care should be taken in handling the following reagents, which should be used under a fume hood.

The liner is first placed in a clean test tube that is at least 4–5 cm longer than the inlet, covered with 25% HCl, and allowed to stand overnight at room temperature or 3–4 hr at a temperature of 60–65°C. The process should be repeated if the acid becomes markedly discolored. The acid is decanted, and the tube and contents are thoroughly rinsed in deionized water and dried at a temperature not exceeding 150°C, or (preferably) under vacuum at a temperature below 100°C; the goal is to produce a nonhydrated fully hydroxlated surface with an absence of strained siloxane bridges [8].

A number of silylation procedures can be employed for silanol deactivation, which constitutes the next step; one very effective method is to draw the neck of the test tube down to a few millimeters and use a disposable pipet inserted through that restriction to add a few drops of diphenyltetramethyl disilazane. A clean pipet is then used to direct a jet of clean dry nitrogen into the tube to displace air, and the tube is flame-sealed. The tube (and contents) are then heated to at least 300°C (higher temperatures are permissable) for at least 3 hr and allowed to cool to room temperature. If the liner is to be stored, it can be left in the sealed tube. The tube should be opened carefully (avoid fumes), the liner rinsed in clean pentane, and installed.

References

1. W. Jennings, "Gas Chromatography with Glass Capillary Columns," 2nd ed. Academic Press, New York, 1980.
2. F. Monari and S. Trestianu, *Proc. Int. Symp. Capillary Chromatography, 5th, 1983,* p. 327. Riva (1983).
3. F. J. Schwende and D. D. Gleason, *J. High Res. Chromatogr.* **8,** 29 (1985).
4. S. A. Mooney, *J. High Res. Chromatogr.* **5,** 507 (1982).
5. E. Guthrie (J & W Scientific, Inc.), personal communication (1985).
6. M. F. Mehran, *J. High Res. Chromatogr.* **9,** 272 (1986).
7. M. M. Thomason, W. Bertsch, P. Apps, and V. Pretorius, *J. High Res. Chromatogr.* **5,** 690 (1982).
8. W. Jennings, "Comparisons of Fused Silica and Other Glass Columns in Gas Chromatography." Huethig, Heidelberg, 1983.

CHAPTER 8

SPECIAL ANALYTICAL TECHNIQUES

8.1 General Considerations

Gas chromatography offers an extremely high potential for separating the components of volatile mixtures, but there are occasions when the method of sample injection or certain other parameters must be modified to achieve a particular goal. This section is not intended as a comprehensive treatment of all the ingenious and specialized methods that have been employed to achieve better results for a given gas chromatographic problem; instead, it presents a highly selective review of a few of the methods that have been used. Sometimes it is simply the complexity of the sample that interferes with the isolation and detection of specific components; heart-cutting techniques, in which a selected group of solutes from one column is redirected to a dissimilar column, can sometimes be used to achieve the necessary separation. On occasions where the goal is the detection of minor components, injection of a massive volume of sample may be required. Large-volume injections can lead to lengthened sample bands at the beginning of the column, with adverse effects on the lengths of eluted solute bands, on peak widths, and on sensitivities. These problems can be especially troublesome in purge and trap methodologies, which usually require some means of refocusing or shortening the sample band deposited on the column. If it assumed that the injection technique has been optimized, the routes to improved separation can usually be reduced to methods of increasing n, α, or k [Eq. (1.23)]. Temperature and column phase ratios are the most general means of manipulating k and were discussed in Chapter 5. Methods of increasing n and α, methods of achieving shorter bands from massive gas injections, and a few selected applications of heart cutting are considered below.

8.2 Flow Stream Switching

Some approaches to be considered here require redirecting the flow stream before or during chromatography. Both mechanical valves and valveless pressure switching devices have been employed for this purpose. The use of fluidic logic elements would seem an ideal way to approach flow stream switching and has been explored by several workers [1–4]. Figure 8.1 illustrates a switch that utilizes two fluidic logic Coanda wall attachment elements. The entering flow stream follows one of the two available routes and is then held to that route by a flow dynamics-generated reduced pressure between the flow stream and the wall. A pressure pulse at that control port momentarily overrides the reduced pressure and switches the stream to the other side of the chamber; the reduced pressure

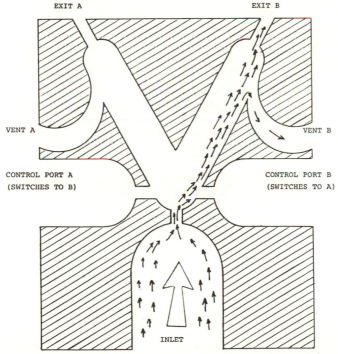

Fig. 8.1. Flow stream switch utilizing two Coanda fluidic logic devices [1–4]. See text for details of operation.

now generated along the second wall will hold the flow stream in this second position until it is canceled by a pressure pulse at that control port. Commercial fluidic switches are available, but in models that are suitable only with flow volumes appreciably larger than those used with all but the largest open tubular columns. A more serious matter is that both fluidic switches and the Deans switch (discussed below) are influenced by differential changes or restrictions in the pressure drops through the alternative flow paths.

Figure 8.2 illustrates a typical Deans switch [5, 15], which utilizes an imposed pressure at the appropriate point to direct the flow stream through the opposite of the two alternate paths. Both the fluidic switch and the Deans switch can be constructed so that only inert materials are in the sample flow stream. On the other hand, temperature shifts (including those encountered in programmed operation) may change the relative pressures in the different legs and inadvertently trigger switching.

Many of the disadvantages cited earlier for mechanical valves [5,6] have been overcome. It is, of course, essential that the flow passages through these devices be consistent with those of the column; excessive volumes associated with the switch would contribute to remixing of partially separated components and to band lengthening. Mechanical valves are quite positive in their switching action, and some models now have internal passages that are consistent with and de-

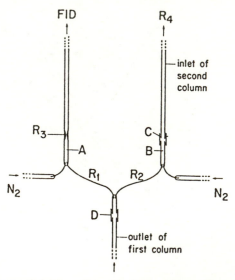

Fig. 8.2. Typical Deans switch [5], configured to direct a column effluent to detection (via R_1) or to a second column (via R_2) [15]. (Reprinted with permission of Heuthig Verlag, Heidelberg, New York.)

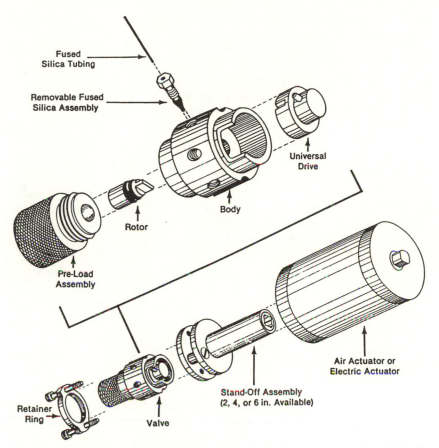

Fig. 8.3. Exploded view of a mechanical valve suitable for flow stream switching [6–8]. (Reproduced by courtesy of the Valco Instrument Co., Inc.)

signed for use with fused silica columns [7,8]. With each passage through the valve, samples come in contact with both the valve body (usually stainless steel) and the rotor material (generally a Teflon-filled polyimide). Valve residence times are usually of the order of milliseconds, and suitability of these valves for a large number of test solutes has been demonstrated [7], but certainly they cannot be considered as totally inert toward all solutes. Among the other disadvantages that have been found for mechanical valves are cold trapping of higher-molecular-weight solutes; such valves usually have an appreciable thermal mass, and the temperature of the valve body usually lags behind that of the oven unless the valve is supplied auxiliary heat [7]. Construction details of a typical mechanical switching valve are shown in Fig. 8.3.

8.3   Recycle Chromatography

Some separations could be improved by employing a separation system that developed very large numbers of theoretical plates; columns of smaller diameter or longer columns are sometimes used in attempts to achieve this goal (see Chapter 5). Several disadvantages, attributable to the higher pressure drops and steeper van Deemter curves, accompany either of these routes.

In studies of the optimum and optimum practical gas velocities of segments of different length from the same glass capillary column, Yabumoto and VandenHeuvel [9] concluded that the OPGV varied inversely with the length of the column. The primary disadvantage of the short column is that it can possess only a limited number of theoretical plates. Recycling a partially resolved fraction through a short column would offer all of the advantages of the short column, while subjecting the solutes to much higher numbers of theoretical plates; several workers have explored this concept [10,11].

Recycling requires transport of the chromatographing band from the column outlet back to the column inlet and is best accomplished by mounting a flow-switching device centrally in the column, i.e., utilizing two short column segments (Fig. 8.4). Imperfections or excess volume in the switching device would cause longer, more dilute bands and encourage remixing of partially separated solutes. A recycle unit employing a prototype mechanical valve connected to two 25-m segments of glass capillary columns has been used to generate over 2,000,000 theoretical plates on a low-k solute (butane) at the optimum carrier gas

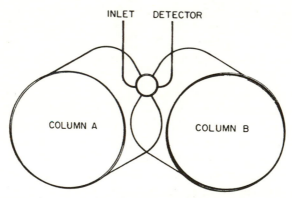

Fig. 8.4. Diagrammatic representation of a recycle unit [14] that permits a short band containing unresolved components to be directed through column segment A to the valve and thence to column segment B; while the band is in segment B, the valve is switched so the band emerging from B is redirected to A. By continuing this process, the band can be recycled through the short column segments some numbers of times before it is allowed to elute to detection. (Reprinted with permission of Huethig Verlag, Heidelberg, New York.)

velocity [12]. When operated at OPGV, a second unit developed over 4000 theoretical plates per second. In essence, the solutes were chromatographed through 400 m of column, but under conditions commensurate with those encountered in a 40-m column. This required 20 repass cycles and a total analysis time of 16 min. A pressure switching unit for recycle chromatography has been suggested but not yet demonstrated [13].

Only a limited section of the chromatographing band can be contained within a column segment; the distance between any two components x and y is

$$d_{x,y} = n_{cyc}[(k_y-k_x)/(k_y+1)]L_{seg} \qquad (8.1)$$

where n_{cyc} is the number of full cycles, L_{seg} the length of one segment ($L_{seg\ 1} = L_{seg\ 2}$), and k_x and k_y represent the partition ratios of the solutes of the two components [14]. Equation (8.1) can also take the form

$$d_{xy}/L_{seg} = n_{cyc}[(k_y-k_x)/(k_y+1)] \qquad (8.2)$$

The limit of the recycle system is reached when $d_{xy} = L_{seg}$; if

$$n_{cyc} > (k_y+1)/(k_y-k_x) \qquad (8.3)$$

some of the chromatographing components must exceed the envelope of the system. Because there is no reason to recycle solutes unless k_x is very nearly equal to k_y, this is not a serious limitation.

A more serious problem in recycle chromatography lies in the coordination of the timing of the valve activation with the chromatography. In the studies cited above, the length of time required for passage of the chromatographing band through one column segment was estimated from an initial injection, and an electronic timer was set to switch the valve pneumatically at that interval. Any error in that set switching interval is cumulative and is ultimately evidenced when a valve switch clips one end from the band and directs it instead to the detector. The ideal moment for valve activation on each pass could be better determined by a nondestructive in-line detector installed between the outlet end of each segment and the valve. Commercially available nondestructive detectors are unsuitable for in-line operation at the flow volumes commensurate with small-diameter open tubular columns. The 530-μm column is sufficiently large to accommodate a microthermistor bead, and this may make recycle chromatography more practical.

Recycle chromatography offers a means of applying a greater number of theoretical plates to unresolved solutes, but there is no advantage to recycling solutes that can be resolved by more ordinary means. The recycle unit should utilize isothermal conditions, and programmed temperatures are preferable for most complex mixtures. Housing the recycle and normal chromatographic units in separating controlled ovens would permit improved flexibility and utility [14].

A band that would otherwise produce an unresolved or multicomponent peak could be redirected to isothermal recycle while the rest of the chromatogram proceeded in a separate unit that could be programmed in the normal manner.

8.4 Multidimensional Chromatography

Forms of multidimensional separations have long been used in gas chromatography; many such applications have been well reviewed [15,16]. The term has occasionally been applied to systems where the effluent from a single column is split to two detectors. Certainly this generates complementary information (e.g., Fig. 9.44), but the additional dimension is in the detection rather than the chromatography. Bertsch [15] reserved the term "two-dimensional gas chromatography" for systems containing two columns of different selectivities operated under conditions that eliminate ambiguities in correlating the solutes eluting from the two columns. The flexible fused silica column has made it very simple, for example, to connect columns coated with dissimilar stationary phases to a single inlet so that the sample is simultaneously injected on the different columns, each of which is connected to a dedicated detector. However, the additional information is of limited value unless it can be unequivocally established that a particular solute is peak x on column A and peak y on column B (e.g., by coinjection of known standards [17]). In general, the utility of this approach varies inversely with the complexity of the sample. Not a great deal of information can be extracted from the fact that 500 solutes exhibit a certain elution pattern on one column and another elution pattern on a different column.

Where peaks from the one column can be correlated with specific peaks from the second column, two-dimensional systems can be employed to establish precise retention indices on two dissimilar stationary phases and have been used for the assignment of specific identifications. It is important to recognize that retention behavior establishes only that a solute cannot be any number of compounds whose retention characteristics on that column under those conditions are known to be demonstrably different. Retentions, in other words, can never prove what a solute is, but only what it is not. If a particular compound is suspected, retention behavior can establish that if it is present in a given sample it must occur as a component of a particular peak.

Assignments based on the agreement of gas chromatographic retentions and mass spectral fragmentation patterns have a high degree of validity because the two probes are orthogonal, or nonredundant [18,19]; i.e., they are based on entirely different properties of the solute. Assignments based on the agreement of retention characteristics on two dissimilar columns, while superior to those employing a single column, are far less reliable because both measurements are based on the volatilities of the solute as affected by its interaction with the different stationary phases.

Flow stream switching is useful in multidimensional chromatography for re-directing from one column a discrete portion of the total chromatographing band (which can be restricted to a single peak) to the second column (heart cutting). Both Deans switches [5,8,20–23] and mechanical valves [6–8] have been used for this purpose. Dual-oven instruments, which permit the two columns to be operated under different temperature profiles, offer distinct advantages [7,8,23–25].

8.5   Specifically Designed Stationary Phases

Some solute mixtures that resist separation on any known stationary phase have been subjected to multidimensional separations, while other efforts in this area have led to the synthesis of well-defined stationary phases, precisely de-signed to maximize the relative retentions of specific groups of solutes.

The first efforts in this direction employed binary stationary phase mixtures and, although often credited to later authors, can be traced to work of Maier and Karpathy [26]. Their approach can be illustrated by considering a very simple hypothetical system consisting of three components, 1, 2, and 3, which yield only two peaks when chromatographed on either of two different stationary phases, A and B. On stationary phase A the two peaks consist of 3 and 1+2, whereas on stationary phase B they consist of 2 and 1+3 (Fig. 8.5, top). Maier and Karpathy established that the distribution constant of a solute in a mixture of two stationary phases amounted to the distribution constant in one phase times the fractional volume of the mixture occupied by that phase, plus the distribution constant in the other phase times the volume fraction of that phase:

$$K_{D(A+B)} = K_{D(A)}\phi_{(A)} + K_{D(B)}\phi_{(B)} \tag{8.4}$$

where $\phi_{(A)}$ and $\phi_{(B)}$ represent the volume fractions of A and B in the binary mixture, respectively. The relationship $K_D = \beta k$ [Eq. (1.14)] permits calculation of distribution constants on columns of known β, or, if β is constant, solute partition ratios can be used instead. These concepts were applied to the simple hypothetical system above in constructing the graph in Fig. 8.5 (bottom), where the partition ratios of each solute in stationary phase A (0 vol % B) are on the left ordinate and values in stationary phase B (100% B) are on the right ordinate.

Equation (8.4) established that the K_D (or at constant β, k) of a solute in a binary mixture is a linear function of the volume fraction of either stationary phase; hence the lines in Fig. 8.5 (bottom) represent the partition ratios of the indicated solute in any binary mixture of A and B ranging from 100% A + 0% B to 0% A + 100% B. This graph makes it apparent that solutes 2 and 3 will exhibit the same K_D (and coelute) in a mixture containing 37% stationary phase B in A. The differentiation of all solute partition ratios will be greatest (and solutes will be most widely dispersed) at mixture compositions where the lines

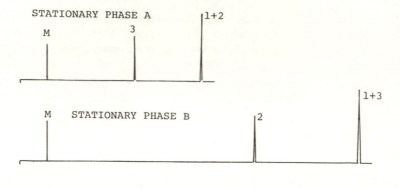

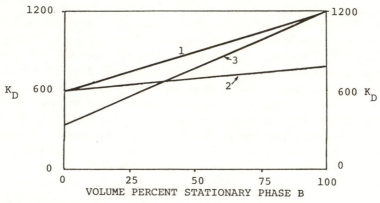

Fig. 8.5. Simplified explanation of the Maier and Karpathy approach to binary stationary phase mixtures [26]. Distribution constants of each solute are calculated from the isothermal chromatogram on stationary phase A and plotted on the left-hand ordinate of the bottom graph; K_D values of each solute on stationary phase B are plotted on the right-hand ordinate. Lines connecting the two points for each solute denote the K_D of that solute for any mixture of stationary phases A and B. See text for discussion.

are most widely spaced. One such point occurs at 22.5 vol % B, where the solute elution order will be 3, 2, and 1 (order of the K_D values); another is at 55 vol % B, where the solute elution order will be 2, 3, and 1. Maier and Karpathy also suggested that the desired proportions of the two stationary phases could be achieved by (1) use of a packing coated with the proper mixture of A and B, (2) blending in a single column the correct proportions of packings that had been separately coated, one with A and one with B, or (3) series coupling of predetermined lengths of two columns, one containing stationary phase A and the other stationary phase B.

Laub and Purnell [27,28] reasoned that the composition of the ideal mixture could be more precisely defined by plotting the relative retentions of each solute pair as a function of the volume percent of one stationary phase in that binary mixture. The lines formed by connecting those points produce a "window diagram" (Fig. 8.6), and the highest "window" occurs at the optimum stationary phase mixture, where the relative retentions of all solutes are maximized.

Ingraham *et al.* [29] used a computerized version of this approach to establish that a mixture of 5 vol % Carbowax 20M in dimethyl polysiloxane would yield the best separation of alcoholic fermentation products whose differentiation had previously required two separations on different columns. These same concepts were then applied to another mixture, whose components were also inseparable on any single column [30]. Figure 8.7 shows a window diagram generated by plotting the relative retentions of every possible solute pair of that mixture as functions of the volume percent of DB-1701 (right axis) in DB-1 (left axis). Windows appear at 25 and at 66 vol % DB-1701. Because the solutes are more retained in DB-1701, solute partition ratios would be larger in the latter window and the magnitude of the $[(k+1)/k]^2$ multiplier of Eq. (1.22) would be significantly smaller (see Table 5.1); a 66 vol % DB-1701 column could therefore achieve the separation with fewer theoretical plates than a 25 vol % DB-1701 column. The appropriate lengths of capillaries coated with the two stationary phases were then coupled to achieve the separation.

It was also noted that because of the diffusivity and velocity gradients engendered by the pressure drop through the coupled columns, the segment at the inlet end influenced the separation more (and the outlet end influenced it less) than would be predicted on the basis of their respective lengths. In other words, the results generated by a coupled column containing equal-length segments separately coated with stationary phases A and B are influenced by the velocity and diffusivity gradients through the column, i.e., on which segment is used at the inlet end. This implies that the earlier generalization (Section 5.3) that the mobile phase velocity is optimized throughout the column by optimizing flow at one point in the column [31,32] is not entirely correct.

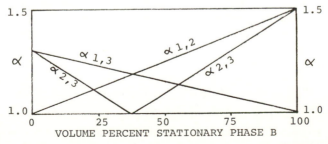

Fig. 8.6. Window diagram for the hypothetical mixture examined in Fig. 8.5 [27,28]. There are two windows, one at 22.5 and one at 55 vol % stationary phase B in A. See text for discussion.

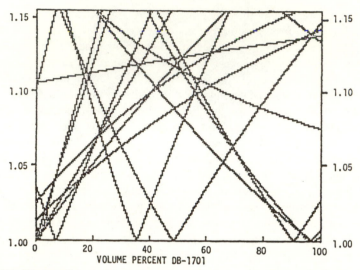

Fig. 8.7. Window diagram for a real mixture [30]; solutes: ethyl hexanoate, ethyl octanoate, geranial, geraniol, limonene, linalyl acetate, myrcene, neral, nonanol, octanal, octyly acetate, and terpinene-4-ol. See text for discussion. (Reprinted with permission of Huethig Verlag, Heidelberg, New York.)

A graphical approach to determining column length fractions required to produce the desired results was suggested by Takeoka *et al.*, but Purnell and Williams [33] were critical of this approach and argued that it was possible to calculate precise correction factors to compensate for the pressure drop-engendered deviations; the validity of the latter argument was not verified experimentally. Krupcik *et al.* [34] concluded that the deviations encountered experimentally did not correspond to calculated values.

Mehran *et al.* applied the coupled-column approach to separate a series of priority pollutants that had resisted separation on any single column [35], and suggested that these deviations could be eliminated by synthesis of a new stationary phase containing the indicated concentrations of the functional groups in the two separate stationary phases. This approach led to the subsequent synthesis of a new stationary phase, DB-1301, specifically designed for the separation of that series of pollutants [36].

8.6   Vapor Samples and Headspace Injections

Headspace injections not only have an appealing simplicity, but also allow the investigator to avoid sample preparation procedures that may engender qualitative and quantitative changes in the composition of the sample [37] and changes that can occur in the interval between sample preparation and sample analysis, even under conditions of low-temperature storage [38]. Minimization of these artifact-related procedures and simplification of the analytical process are strong incentives for the use of direct sample injections whenever possible.

On the other hand, headspace injections are normally dominated by the more volatile components of the sample, and the detection of higher-boiling and/or trace components has been difficult. Techniques such as purge and trap [39] were developed largely to permit prechromatographic isolation of detectable quantities of trace components from volumes of vapor too massive for their direct injection. Although these methods extend the utility of vapor phase analysis, problems related to inadequate retention of some components by the trapping substrate ("breakthrough," usually of the more volatile constituents) and inadequate recovery of other components from the trapping substrate (sometimes resulting from posttrapping degradation on the substrate) have also been reported (see, e.g. [37]).

The primary benefit of purge and trap techniques is that larger volumes of vapor can be sampled; the limits of detection can be considerably extended, provided the volatiles entrained in that sample can be recovered from the gaseous matrix and introduced onto the chromatographic column as a short starting band. Toward the latter goal, the carrier gas stream is normally employed to discharge the contents of the loaded trap into the column; the trap is usually heated during this step, while the column (or a portion thereof) is chilled.

By using a syringe with a very fine needle, it is also possible to eliminate the intermediate trapping and discharge steps and deposit the vapor sample directly inside the chilled column [40]. Fused silica needle syringes have been used for this purpose quite successfully. The method has also been used for the detection of halocarbons in water, with detection limits reportedly extending into the parts-per-trillion range [41]. Work now in progress indicates that by injecting headspace samples as large as 10 cm^3, sensitivities for even higher-boiling and strongly polar solutes in aqueous samples can be greatly enhanced. Solutes contained in the vapor sample must be given sufficient opportunity to cold trap as a focused band in the first chilled section of the column. To avoid an elongated sample band and broader peaks, lower injection speeds should be employed for samples dominated by lower-boiling solutes, or where the column is chilled with carbon dioxide as opposed to liquid nitrogen. The speed of injection should also be decreased as the sample size is increased.

A major limitation of these procedures is that they require considerable manual manipulation and dexterity. Encouraging results have been obtained in preliminary attempts to simplify the introduction and removal of coolant in such a way as to enhance shortened bands (Fig. 8.8), which may make possible full automation of this procedure for vapor phase sampling.

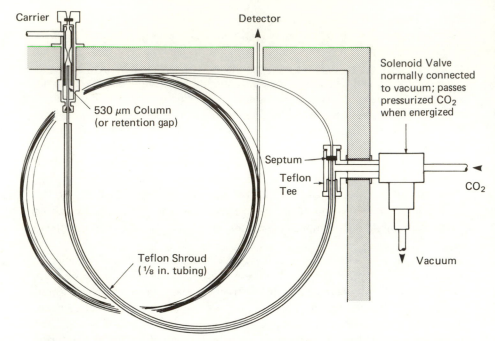

Fig. 8.8. Schematic representation of a device for achieving focused bands with on-column headspace injections. By introducing the coolant counter to and the heat in the same direction as the carrier gas flow, the band is shortened during both the trapping and injection steps. (Drawing by R. Masterson. Reproduced with permission of the copyright holder, J&W Scientific, Inc.)

References

1. R. L. Wade and S. P. Cram, *Anal. Chem.* **44,** 131 (1972).
2. S. P. Cram and S. N. Chesler, *J. Chromatogr.* **99,** 267 (1974).
3. G. Gaspar, P. Arpino, and G. Guichion, *J. Chromatogr. Sci.* **15,** 256 (1977).
4. R. Annino, M. F. Gonnord, and G. Guichion, *Anal. Chem.* **51,** 379 (1979).
5. D. R. Deans, *Chromatographia* **1,** 18 (1968). G. Schomburg, F. Weeke, F. Mueller, and M. Oreans, *Chromatographia* **16,** 87 (1983).
6. R. J. Miller, S. D. Sterns, and R. R. Freeman, *J. High Res. Chromatogr.* **2,** 55 (1979).
7. W. Jennings, *J. Chromatogr. Sci.* **22,** 129 (1984).
8. B. M. Gordon, C. E. Rix, and M. F. Borgerding, *J. Chromatogr. Sci.* **23,** 1 (1985).
9. K. Yabumoto and W. J. A. VandenHeuvel, *J. Chromatogr.* **140,** 197 (1977).
10. D. Dedford (Phillips Petroleum Co.), U.S. Patent 3,455,090 (1969).
11. A. M. Ried, *J. Chromatogr. Sci.* **14,** 203 (1976).
12. W. Jennings, J. A. Settlage, R. J. Miller, and O. G. Raabe, *J. Chromatogr.* **186,** 189 (1979).
13. C. D. Chriswell, *J. High Res. Chromatogr.* **5,** 210 (1982).
14. W. Jennings, J. A. Settlage, and R. J. Miller, *J. High Res. Chromatogr.* **2,** 441 (1979).
15. W. Bertsch, *J. High Res. Chromatogr.* **1,** 187, 289 (1978).
16. D. R. Deans, *J. Chromatogr.* **203,** 19 (1981).
17. C. W. Wright, *J. High Res. Chromatogr.* **7,** 83 (1984).

18. D. H. Freeman, *Anal. Chem.* **53,** 2 (1981).

19. J. C. Giddings, *Anal. Chem.* **39,** 1027 (1967).

20. G. Schomburg, H. Husmann, and F. Weeke, *J. Chromatogr.* **112,** 205 (1975).

21. R. J. Phillips, K. A. Knauss, and R. R. Freeman, *J. High Res. Chromatogr.,* **5,** 546 (1982).

22. G. Schomburg, F. Weeke, F. Mueller, and M. Oreans, *Chromatographia* **16,** 87 (1983).

23. G. Schomburg, E. Bastian, H. Behlau, H. Husmann, and F. Weeke, *J. High Res. Chromatogr.* **7,** 4 (1984).

24. "Two-Oven Gas Chromatograph," Prod. Bull. IBM Instruments, Danbury, Connecticut.

25. Applications Notes 282, 297. Siemens AG, Karlsruhe, Germany; see also Application News 1, 2, 3. ES Industries, Marlton, New Jersey.

26. H. J. Maier and O. C. Karpathy, *J. Chromatogr.* **8,** 308 (1962).

27. R. J. Laub and J. H. Purnell, *J. Chromatogr.* **112,** 71 (1975).

28. R. J. Laub and J. H. Purnell, *Anal. Chem.* **48,** 799 (1976).

29. D. F. Ingraham, C. F. Shoemaker, and W. Jennings, *J. Chromatogr.* **239,** 39 (1982).

30. G. Takeoka, H. M. Richard, M. Mehran, and W. Jennings, *J. High Res. Chromatogr.* **6,** 145 (1983).

31. J. C. Giddings, *Anal. Chem.* **36,** 741 (1964).

32. J. C. Sternberg, *Anal. Chem.* **36,** 921 (1964).

33. J. H. Purnell and P. S. Williams, *J. High Res. Chromatogr.* **6,** 799 (1983).

34. J. Krupcik, G. Guichion, and J. M. Schmitter, *J. Chromatogr.* **213,** 189 (1981).

35. M. F. Mehran, W. J. Cooper, and W. Jennings, *J. High Res. Chromatogr.* **7,** 215 (1984).

36. M. F. Mehran, W. J. Cooper, R. Lautamo, R. R. Freeman, and W. Jennings, *J. High Res. Chromatogr.* **8,** 715 (1985).

37. W. Jennings and A. Rapp, "Sample Preparation for Gas Chromatographic Analysis." Huethig, Heidelberg, 1983.

38. G. Takeoka, M. Guentert, S. L. Smith, and W. Jennings, *J. Agric. Food Chem.* (1986) (in press).

39. T. A. Bellar and J. J. Lichtenberg, *J. Am. Water Works Assoc.* **66,** 739 (1974).

40. G. Takeoka and W. Jennings, *J. Chromatogr. Sci.* **22,** 177 (1984).

41. M. F. Mehran, W. J. Cooper, M. Mehran, and W. Jennings, *J. Chromatogr. Sci.* **24** (1986) (in press).

CHAPTER 9
SELECTED APPLICATIONS

9.1   General Considerations

Any attempt to create a compilation of selected applications intended as a guideline must be selective; in consequence, that compilation can almost always be faulted by individual chromatographers for the omission of examples that they consider critical to their specific fields. On the other hand, some of the included examples can be criticized because developments in gas chromatography continue to occur at a rapid pace; in many cases it is now possible to generate results that are superior to the examples used as illustrations. Developments that have made those improvements possible include (1) better methods of sample pretreatment and storage, (2) improvements in commercial instrumentation, including superior inlet devices that take into account our newer understandings of the injection process, and (3) the availability of better deactivated columns in a wider range of diameters and coated with a greater variety of stationary phase types and film thicknesses. Some of the chromatograms used as examples in this chapter could now be improved on because of developments of this type that have occurred since those chromatograms were originally obtained. But although it is true that today's analyst can usually reap the advantages of at least another year or two of developmental improvements, many of the examples cited can still be useful as points of departure.

Application guidelines for stationary phase selection used in catalogs of suppliers such as Supelco, Applied Science (now Alltech), Analabs, and others can also be useful, and information from those sources forms the basis of the tabulations inserted at appropriate points in this chapter. Intercomparisons are complicated by the fact that columns coated with a particular stationary phase carry different names and codes depending on the manufacturer. In actuality, those columns may in fact be different; most manufacturers purchase stationary phases per se, some refine the purchased preparations, and others synthesize their stationary phases in-house. Qualitative differences are common between different lots of tubing and can also occur in different sections of the same lot. Variations in the deactivation processes, in coating procedures, and in the level of quality control all influence the ultimate behavior of the finished column. Stationary phase citations used here are not intended as endorsements of any particular brand; stationary phase "equivalents" are indicated in Table 9.1. At the same time, it must be recognized that columns containing stationary phases of a given "type" but produced by different manufacturers are not necessarily equivalent. Most stationary phases are obtained from polymer suppliers whose chromatographic sales represent a minuscule portion of a market that is primarily directed to lubricants or detergent chemicals; a column manufacturer resorting to in-house synthesis of stationary phases takes this more expensive route in an effort to increase column quality and reproducibility. In addition, there are a few cases (e.g., the DB-1301 and DB-624 of J&W; the Al_2O_3 PLOT columns of Chrompak) where no equivalent phases exist at the time of this writing.

Another approach to the classification of applications in gas chromatography

TABLE 9.1

"Equivalent" Stationary Phases

Type	J & W	HP	Quadrex	Chrompak	Supelco	SGE	RSL
1[a]	DB-1	Ultra-1	1	CP-Sil-1CB	SPB-1	BP 1	150
5[b]	DB-5	Ultra-2	2	CP-Sil-8CB	SPB-5	BP 5	200
17[c]	DB-17	HP-17	17		SP-2250		300
210[d]	DB-210						400
225[e]	DB-225		225	CP-Sil-43CB		BP 15	500
275[f]	DL-2330			CP-Sil-84	SP-2330		
1701[g]	DB-1701		1701	CP-Sil-19CB		BP 10	
PEG[h]	DB-Wax		CW	CP-Wax-51CB	Supelco wax-10	BP 20	

[a]100% dimethyl polysiloxane (SE-30, OV-101).

[b]5% phenyl, 1% vinyl methyl polysiloxane (SE-54).

[c]50% phenyl methyl polysiloxane (OV-17, SP-2250).

[d]50% trifluoropropyl methyl polysiloxane (OV-210, QF-1, SP-2401).

[e]25% phenyl, 25% cyanopropyl methyl polysiloxane (OV-225, XE-60).

[f]70–80% cyanopropyl methyl polysiloxane (OV-275, SP-2340, SP-2330).

[g]6% phenyl, 6% cyanopropyl methyl polysiloxane (OV-1701).

[h]The various "bonded" polyethylene glycol stationary phases are markedly different in (1) maximum and (2) minimum operating temperature limits and (3) their solubilities in (and compatibility with injections containing) water and low-molecular-weight alcohols.

is to divide pertinent data in the literature into subject matter areas; although some overlap is inevitable, a serious degree of redundancy can be avoided by assignments to the four major categories: (1) food, flavor, and fragrance; (2) petroleum and chemical; (3) environmental; and (4) biological and medical. Natural products and pheromones are most logically included in the first group, as are chemicals of major significance in the food processing industries (e.g., monitoring of ethylene dibromide and nitrosamine in foods and diethylene glycol in wines). Fatty acids have been considered under both the food and biological headings. Analyses concerned with synthetic fuels, coal, and shale are considered under the second section. Examples of forensic applications such as the detection of accelerants in arson residues appear in the second section, while measurements of blood alcohol and drugs of abuse are included in the fourth (biological) section. The third section considers the detection of contaminants in air, water, and soil, ranging from halomethanes in water to dioxins in soil. Pesticide analyses appear in both sections 1 and 4; pesticide contamination is usually a matter of environmental concern, but different procedures may be required for the detection of residual pesticides in foods.

Separation of enantiomers can be of interest in all of the above fields; the protein amino acids are considered under food, while the other restricted examples cited here were arbitrarily assigned to the biological area.

9.2   Food, Flavor, and Fragrance Applications

Column recommendations for food, flavor, and fragrance applications that can be gleaned from suppliers' catalogs could be classified as shown in Table 9.2.

TABLE 9.2

Columns Commonly Used for Food, Flavor, and Fragrance Applications

Solutes[a]	Stationary phase	Column type[b]
Volatile fatty acids in water (C_2 to C_5)	10% SP-1000 (or 10% SP-1200 or 15% SP-1220)/1%H_3PO_4; 10% AT-1400 or 10% AT-1000; Carbopak C/0.3% Carbowax 20M/1% H_3PO_4	PEGa
Fatty acids, bacterial	3% SP-2100 (or OV-101); Silar 5CP (or 10CP); DEGS or OV-225	1, 5, 225, 275
Amino acids (trimethylsilyl derivatives) (n-butyl TFA)	OV-11 OV-17, OV-210	17 210
Carbohydrates (trimethylsilyl or alditol acetate)	2330, 2340	225, 275

[a]Abbreviations: TMS, trimethylsilyl; TFA, trifluoroacetate derivative.
[b]See Table 9.1 for "equivalent" columns.

Most of the objective approaches in flavor research employ gas chromatography, but sensory correlations are required before any flavor significance can be assigned to gas chromatographic patterns. The analyses may be complicated by the fact that the volatiles of interest usually occur at very low concentrations, and they are often dispersed through a matrix containing materials which would, if introduced onto the column, shorten its useful life to a considerable degree; a sample preparation step may be essential. Chromatograms can be extremely complex and are often influenced by the sample preparation methods employed.

Figure 9.1 shows chromatograms of black pepper produced by (top to bottom) headspace injection, simultaneous steam distillation/extraction, and Soxhlet extractions with dichloromethane and liquid carbon dioxide [1]. The headspace

Fig. 9.1. Chromatograms of black pepper volatiles produced by (top to bottom) (1) direct injection of 200 μl of headspace vapor, (2) simultaneous steam distillation extraction (Nickerson–Likens), and Soxhlet extractions with (3) dichloromethane and (4) high-pressure liquid carbon dioxide [1]. Column, 30 m × 0.25 mm, 0.25-μm film of a bonded dimethyl polysiloxane; 40–280°C at 4°/min. Headspace injection on-column, all other injections split 1 : 100. (Reproduced by permission of the American Chemical Society.)

injection accentuates the extremely volatile constituents and discriminates against higher-boiling components; Soxhlet extraction discriminates against the more volatile solutes and accentuates the higher-boiling substances; results obtained by simultaneous steam distillation/extraction lie between the two. Soxhlet extraction with liquid carbon dioxide is especially useful for fragile essences that could be damaged by heat; Fig. 9.2 illustrates the results of a flower extraction.

Gas chromatography is widely used in the analysis of the essential oils, where the complexity of the sample can offer a very real challenge. Many of these products are both fragile and commercially important; chromatographic separations are employed to quantitate specific compounds that may be indicative of positive or negative quality notes in the oil, and to detect the clandestine addition of addends such as antioxidants or less expensive oil components used as diluents. Lemon oil, which has a long history as an important oil of commerce, can serve as an example. An early form of lemon oil adulteration was the addition of turpentine [2], but modern analytical methods have led instead to dilution with less expensive but normal constituents of lemon oil (e.g., benzyl alcohol [3,4]). Hence gas chromatography is used not only to establish that the content of such materials is within normal ranges but also to measure other components whose concentrations are known to increase as the oil is abused (e.g., p-cymene, myrcene), to quantitate the more desirable constituents (e.g., citral), and to detect preservatives such as butylated hydroxytoluene and butylated hydroxyanisole (Fig. 9.3 [5]). The complexity of the oil is such that no single column is able to resolve completely all components. One major commercial facility performs its lemon oil analyses on high-methyl polymethylsiloxane columns with the realization that other stationary phases, while yielding better overall separation, exhibit shorter lifetimes; the longer-life methyl silicone column separates all components regarded as critical. The same reasoning can be applied to the analysis of a number of other complex mixtures; Figs. 9.4 and 9.5 show chromatograms of a peppermint oil and of a cologne on such columns.

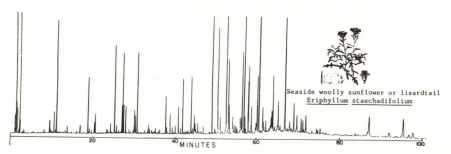

Seaside woolly sunflower or lizardtail
Eriphyllum staechadifolium

Fig. 9.2. Chromatogram of an essence of *Eriophyllum staechadifolium* (seaside woolly sunflower, also known as lizardtail), produced by high-pressure liquid carbon dioxide Soxhlet extraction [1]. Chromatographic conditions as in Fig. 9.1. (Reproduced by permission of the American Chemical Society.)

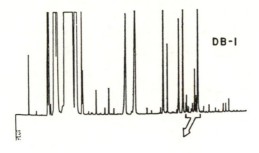

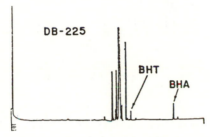

Fig. 9.3. Heart cutting applied to the detection of butylated hydroxytoluene (BHT) and butylated hydroxyanisole (BHA) in a commercial lemon oil. Retention of the two antioxidants on each of the two dissimilar columns was first established by injections of the pure materials; a narrow "window" that would embrace those solutes was then redirected from the 30 m × 0.32 mm dimethylpolysiloxane column to the 30 m × 0.32 mm methylphenylcyanopropyl siloxane column [5]. Chromatographic conditions: top, 85–95°C at 1°/min, 2°/min to 110°C, 8°/min to 220°C; bottom, 30°C through cut, rapid ballistic rise to 70°C, 10°/min to 220°C. (Adapted from the *Journal of Chromatographic Science*, volume 22, 1984, p. 177, with permission of the copyright holder, Preston Publications, Inc.)

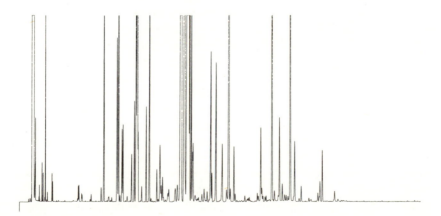

Fig. 9.4. Peppermint oil on a 30 m × 0.32 mm, 95% dimethyl siloxane column. On-column injection of 0.5 μl, 40°C for 0.5 min, 40–60°C at 60°/min, and to 250°C at 3°/min.

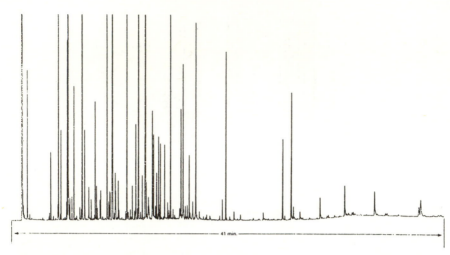

Fig. 9.5. A commercial cologne; 1 μl split 1 : 100 to a 30 m × 0.25 mm, 95% dimethyl siloxane column; 60°C for 2 min, 8°/min to 300°C.

In their studies of the volatile constituents of black tea, Mick and Schreier [6] subjected tea leaf infusions to vacuum steam distillation, solvent extraction, and fractionation on silica gel. Figure 9.6 illustrates the complexity of three such fractions. The volatiles of Rooibos tea were studied by Habu *et al.* [7]; Fig. 9.7 compares chromatograms of tea leaf headspace volatiles isolated by Tenax trapping with those of a vacuum steam distillate [8]. Capillary gas chromatography and GC/MS were used to identify almost 100 compounds in the vacuum steam volatile oil and 218 compounds in the headspace of the dry leaves.

Figure 9.8 shows a chromatogram typical of that produced by headspace injection of vapors overlying ripening banana [9]. Chromatograms typical of those produced by injections of headspace samples from smoked meat products are shown in Fig. 9.9 [10].

Injection of 500 μl of a wine headspace produced the chromatogram shown in Fig. 9.10 [11]; the first major peak is probably acetaldehyde, followed by methyl acetate. Note that the PEG-type stationary phase does not normally differentiate 2-methylbutanol from 3-methylbutanol; a high-methyl polysiloxane-type stationary phase achieves slightly better separation of these two solutes. With wine as with other materials, chromatograms of liquid–liquid extraction essences are usually more complex than those produced by headspace injection (Fig. 9.11) [11].

An adulteration problem affecting some European wines has assumed great importance at the time of this writing: unreasonable quantities of diethylene glycol were initially found in certain Austrian wines which were exported to other countries and used for blending. The material is, of course, toxic and has

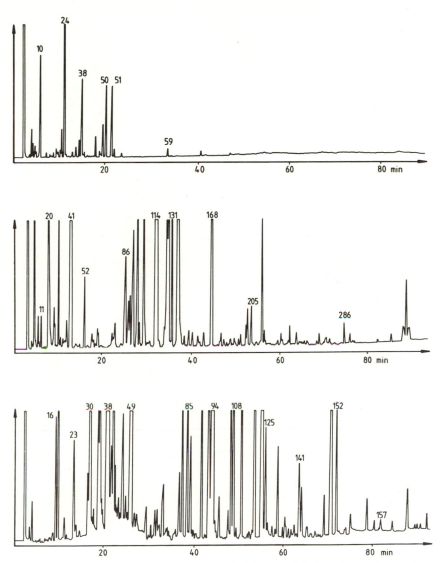

Fig. 9.6. Chromatograms of black tea volatiles [6]. A hot-water infusion of *Camellia sinensis* was subjected to vacuum steam distillation, and the concentrated pentane–dichloromethane extract fractionated by gradient elution from a silica gel column. The fraction shown in the top chromatogram was least polar and eluted with pentane; the fractions in the center and bottom chromatograms eluted with increasing concentrations of ethyl ether. Column, 30 m × 0.32 mm, polyethylene glycol, 5 min at 60°C, 2°/min to 240°C. (Reproduced by permission of the American Chemical Society.)

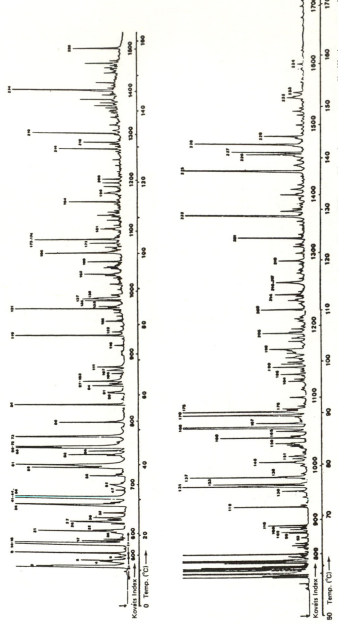

Fig. 9.7. Chromatograms of Rooibos tea [7]. Top, headspace vapors trapped on a Tenax GC cartridge; bottom, steam distillation–extraction essence. Column, 60 m × 0.32 mm, bonded dimethyl polysiloxane; 50–250°C at 3°/min. (Reproduced by permission of the American Chemical Society.)

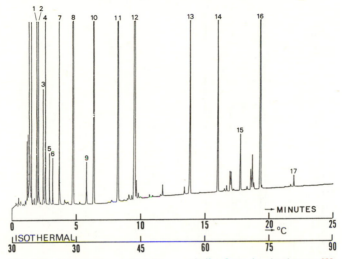

Fig. 9.8. Chromatogram of headspace volatiles from ripening banana [9].

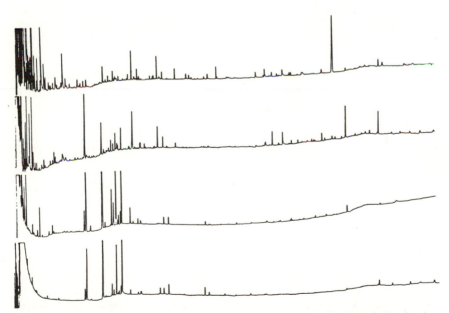

Fig. 9.9. Comparison of chromatograms produced by a commercial "liquid smoke" and various smoked meat products [10]. Chromatograms (from top), 1 ml liquid smoke and 10 ml headspace injections of smoked bacon, Landjaeger, and Berliner Schinkenknacker (bottom).

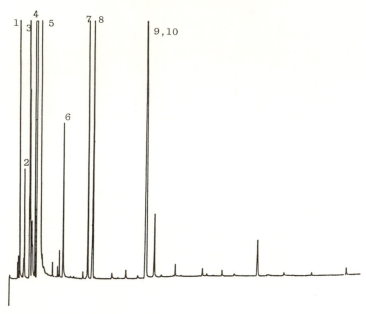

Fig. 9.10. Chromatogram produced by injection of 500 μl of Muscat wine headspace on a 30 m × 0.32 mm, bonded PEG column [11]; 30°C for 2 min, 2°/min to 40°C, 4°/min to 170°C [11]. Peak assignments: (1) acetaldehyde, (2) methanol, (3) ethyl acetate, (4) ethanol, (5) propanol, (6) 2-methyl-1-propanol, (7) butanol, (8) 2-pentanol, (9) 2-methyl-1-butanol, (10) 3-methyl-1-butanol.

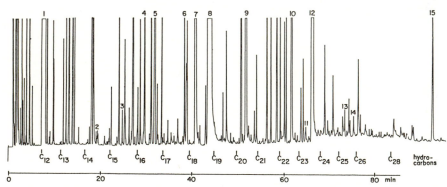

Fig. 9.11. Chromatogram of a dichloromethane–Freon 11 extract of a California Riesling on a 30 m × 0.32 mm, bonded PEG column [11]; 40°C for 1 min, 2°/min to 200°C. Retentions of the *n*-paraffin hydrocarbons as indicated. Solutes: (1) 2-methyl-1-butanol and 3-methyl-1-butanol, (2) furfural, (3) linalool, (4) ethyl decanoate, (5) diethyl succinate, (6) 2-phenethyl acetate, (7) hexanoic acid, (8) 2-phenethanol, (9) octanoic acid, (10) decanoic acid, (11) dihydroxyhexanoic acid gamma lactone, (12) monoethyl succinate, (13) N-(2-phenethyl)acetamide, (14) ethyl pyroglutamate, (15) 2-(4-hydroxyphenyl)ethanol.

since been isolated from other wines; the contaminant has been found in levels as high as 60 g/liter [12]. The complexity of the normal wine chromatogram complicates the detection of this material unless the assignment is based on GC/MS or correlation with some other nonredundant probe. Kaiser [12] demonstrated the detection of glycol in spiked wine by multidimensional gas chromatography; he also suggested that some of the glycol "adulterant," particularly where it occurs at very low levels, may have been produced by natural fermentation processes.

Markides *et al.* [13] pointed out that separation of the cis/trans isomers of fatty acid methyl esters (FAME) can be readily achieved on a polar stationary phase, whereas both a polar phase and an efficient column are required to separate FAME mixtures containing positional and geometric isomers. Separation of a FAME mixture from a herring liver extract on an experimental column prepared by that group is shown in Fig. 9.12. Figure 9.13 illustrates separation of a standard FAME test mixture on a commercial fused silica column coated with a partially cross-linked, surface-bonded, 10% phenyl–90% cyanopropyl polysiloxane stationary phase, and Fig. 9.14 shows the methyl esters of bacterial fatty acids on the same stationary phase.

The chromatogram produced by a racemic mixture of derivatized protein amino acids on a Chirasil-Val column is shown in Fig. 9.15 [14]; a similar separation on a stationary phase composed of polyethylene glycol Carbowax 20M cross-linked with a chiral phase is shown in Fig. 9.16 [15].

Gas chromatographic analysis of the mycotoxins has been handicapped by decomposition of the derivatives during analysis. Kientz and Verweij studied the efficacy of several derivatizing reagents for the type A and type B trichothecenes and reported that such decompositions were largely attributable to the presence

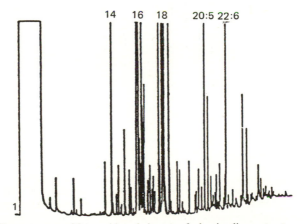

Fig. 9.12. Chromatogram of fatty acid methyl esters of a herring liver extract on a 20-m column coated with a 75% cyanopropyl silicone [13].

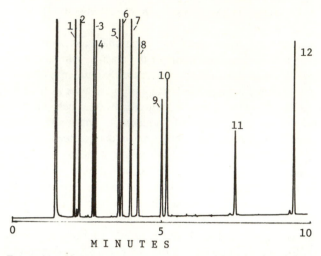

Fig. 9.13. Fatty acid methyl esters, standard mixture, 0.2 μl split 1 : 200 on a 30 m × 0.25 mm column, 0.25-μm film of a 10% phenyl, 90% cyanopropyl stationary phase (DL-2330). Conditions, 6 min at 190°C, 10°/min to 250°C. Solutes: (1) 14:0, (2) 14:1, (3) *trans*-16:1, (4) *cis*-16:1, (5) *trans*-18:1, (6) *cis*-18:1, (7) *trans,trans*-18:2, (8) *cis,cis*-18:2, (9) 20:1, (10) 18:3, (11) 22:1, (12) 24:1.

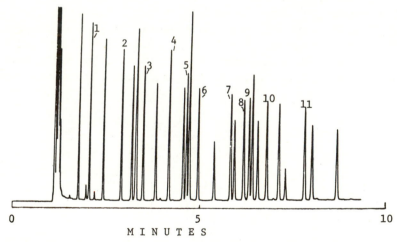

Fig. 9.14. Fatty acid methyl esters (Supelco No. 4-7080 bacterial acid methyl esters). Column, 30 m × 0.25 mm, 0.25-μm film of 10% phenyl, 90% cyanopropyl stationary phase (DL-2330). Conditions, 0.5 μl split 1 : 100; 155–270°C at 6°/min. Numbered peaks: (1) 12:0, (2) 13:0, (3) 14:0, (4) 16:0, (5) *cis*-16:1, (6) 17:0, (7) 18:0, (8) *trans*-18:1, (9) *cis*-18:1, (10) 19:0, (11) 20:0. Sample also contained: methyl undecanoate; 2-OH decanoate; 2-OH dodecanoate; 3-OH dodecanoate; 13-methyl tetradecanoate; 12-methyl tetradecanoate; 2-OH tetradecanoate; 3-OH tetradecanoate, 14-methyl pentadecanoate, 15-methyl hexadecanoate; 14-methyl hexadecanoate; *cis*-9,10-methylene hexadecanoate; 2-OH hexadecanoate, *cis*-9,12-octadecenoate; and *cis*-9,10-methylene octadecanoate.

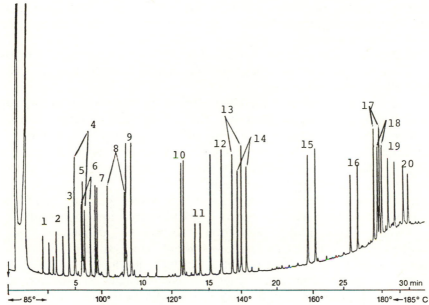

Fig. 9.15. Separation of protein amino acids as N-(O,S)-pentafluoropropionyl isopropyl esters on Chirasil-Val 8 [14]. Solutes: (1) alanine, (2) valine, (3) threonine, (4) α-isoleucine, (5) glycine, (6) isoleucine, (7) proline, (8) leucine, (9) serine, (10) aspartic acid, (11) cysteine, (12) methionine, (13) phenylalanine, (14) glutamine, (15) tyrosine, (16) ornithine, (17) lysine, (18) histidine, (19) argenine, (20) tryptophan. In each case, the D isomer elutes first.

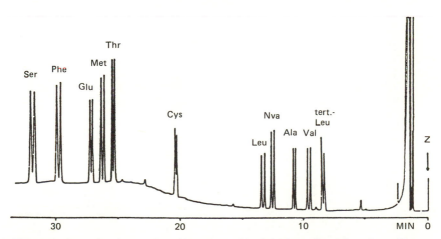

Fig. 9.16. Racemic mixture of N-trifluoroacetamide derivatives of protein amino acid enantiomers on an experimental stationary phase, Carbowax 20M cross-linked with L-valine-(S)-α-phenylethylamide [15].

of excess reagent during the analysis [16]. Figure 9.17 shows a chromatogram of mycotoxins derivatized with the volatile trifluoroacetic anhydride in the presence of sodium bicarbonate.

Headspace sampling can be very effective in the determination of residual solvents in foodstuffs and has, for example, been used to detect solvent residues in decaffeinated instant coffee [17].

Some food packaging (and handling) materials possess a potential for contamination that must also be addressed. That potential may be related to trace amounts of residual solvents used in the manufacture of food packaging films, to the presence of extractable plasticizers (e.g., 2-ethylhexanol in plastic hose and tubing), or to degradation products from the packaging materials. Plastic bottles of polyethylene terephthalate, intended for use as beverage containers and stored capped to preserve their cleanliness, have been found to contain significant quantities of acetaldehyde [18]. A related problem occurs in the use of plasticware with convenience foods. Some frozen "TV dinners" are purchased "microwave-ready" with the edible materials artfully arranged on plastic dishes. Heat is transferred to the plastic in contact with the heated food, which would be expected to encourage migration of residual monomers and catalysts from the plastic into the foodstuff [19]. Figure 9.18 compares the total ion current chromatograms of headspace concentrates from two batches of polyvinyl chloride (PVC), bags from which are destined for the storage of intravenous solutions [20].

Ethylene dibromide (EDB) has been widely used to inhibit postprocessing infestation by weevils and other insects in packaged foods such as grain products

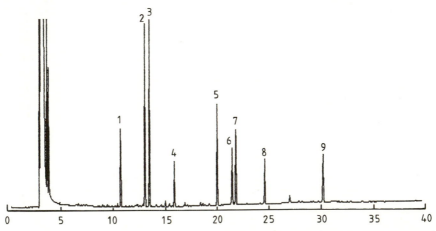

Fig. 9.17. Mycotoxins (trichothecenes) derivatized with trifluoroacetamide in the presence of sodium bicarbonate [16].

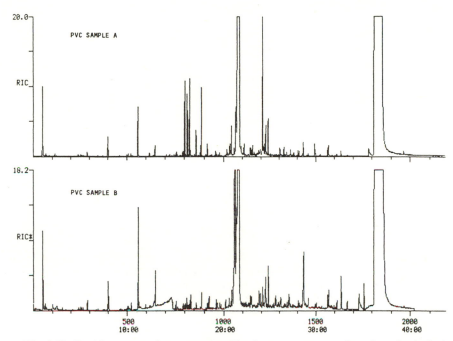

Fig. 9.18. Total ion current chromatograms of headspace concentrates of two lots of polyvinyl chloride [20]. Column, 24 m × 0.21 mm type 5 (SE-54), directly interfaced with the ion source of a Jeol MS-300 mass spectrometer. Solutes identified (scan number in parentheses): Top, (559) 2-ethyl hexanol, (1055) butylated hydroxyanisole, (1071) butylated hydroxytoluene, (1850–1860) bis(2-ethylhexyl)phthalate. Bottom, (560) 2-ethyl hexanol, (600–730) 1,2,3-propanetriol, (1056) butylated hydroxyanisole, (1071) butylated hydroxytoluene, (1815–1860) bis(2-ethylhexyl)phthalate.

and cake mixes. It was reasoned that the material would be expelled from the foodstuff during subsequent cooking, but the demonstrated carcinogenicity of higher levels of EDB triggered concern and a number of legislative acts banning that chemical. Because of the complexity of most samples isolated from food, unambiguous separation of something like EDB can be a problem. Section 8.5 describes techniques employed to design stationary phases for specific applications based on maximizing the relative retentions of all solutes and how those techniques were applied to the synthesis of a new stationary phase tailored to halocarbon separations (see also [21]). The DB-624 column, described in Section 9.5, employs a thicker film of this stationary phase to achieve separation of a series of volatile priority pollutants by optimization of both relative retentions (α) and partition ratios (k). This column is also capable of achieving the separation of ethylene dibromide from several solutes with which it is prone to coelute (Fig. 9.19 [22]).

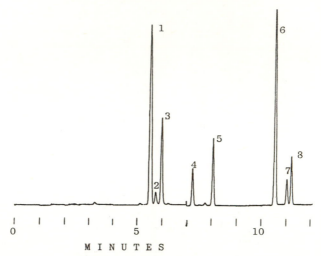

Fig. 9.19. Detection of ethylene dibromide (EDB) [22]. Column, 30 m × 0.53 mm, DB-624; 35–120°C at 5°/min; electron capture detection. Solutes: (1) chloroform, (2) 1,1,1-trichloroethane, (3) carbon tetrachloride, (4) trichloroethylene, (5) bromodichloromethane, (6) tetrachloroethylene, (7) chlorodibromomethane, (8) ethylene dibromide. (Reprinted with permission of the copyright holder, J&W Scientific, Inc.)

9.3   Petroleum- and Chemical-Related Applications

Petroleum-related analyses range from the separation of light gas mixtures to the characterization of heavy crude oils to oil shales and materials produced in coal liquefaction. Particularly with the naturally occurring materials, the dominant solutes are hydrocarbons.

Differentiations of the paraffinic, olefinic, and aromatic hydrocarbons are normal goals, and detection of sulfur- and/or nitrogen-containing components is sometimes required. Hydrocarbon preparations derived from shale oil or coal liquefaction can be extremely complex and can also include a variety of other functional groups. Multidimensional chromatography may be required for the analysis of these complex mixtures; it is also useful for the estimation of certain octane-boosting additives in automotive fuels (e.g., [5]).

Packed columns are still widely used for some separations of light hydrocarbons, fixed gases, low-molecular-weight sulfur- and nitrogen-containing compounds, and low-molecular-weight alcohols and halocarbons. Packings for these applications usually consist of molecular sieve, alumina, or one of the porous polymers (Chromosorb 104, Chromosorb 106, Porapak N).

Interactions between the stationary phase and hydrocarbon-type solutes are generally limited to dispersion forces (Chapter 4); the methyl polysiloxane stationary phases, whose interaction potentials are limited to dispersion, are well suited to the separation of these compounds. At low temperatures, methyl poly-siloxanes (like other substances) lose their fluid characteristics and behave more like conventional solids. Cross-linking slows this low-temperature transition, and higher diffusivities persist for some time at subambient temperatures. Held at low temperatures for extended periods, they eventually exhibit solidlike properties and chromatography suffers; a brief exposure to higher temperature restores

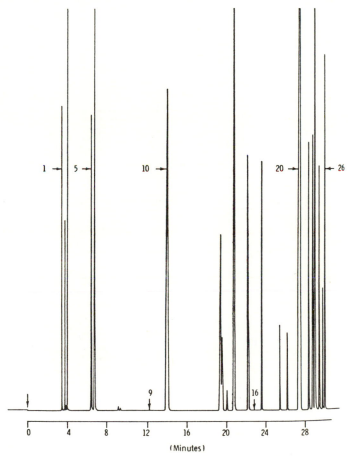

Fig. 9.20. Separation of C_1–C_5 hydrocarbons [23]. Column, 60 m × 0.25 mm, bonded poly-methylsiloxane, d_f 1.0 μm; −40°C for 15 min, to 100°C at 10°/min. Solutes: (1) methane, (2) ethylene, (3) acetylene, (4) ethane, (5) propylene, (6) propane, (7) propadiene, (8) methylacetylene, (9) cyclopropane, (10) isobutane, (11) isobutylene, (12) 1-butene, (13) 1,3-butadiene, (14) *n*-butane, (15) *trans*-2-butene, (16) methylcyclopropane, (17) *cis*-2-butene, (18) 1,2-butadiene, (19) 3-methyl-1-butene, (20) *n*-pentane, (21) 1-pentene, (22) 2-methyl-1-butene, (23) *n*-pentane, (24) *trans*-2-butene, (25) *cis*-2-pentene, (26) 2-methyl-2-butene.

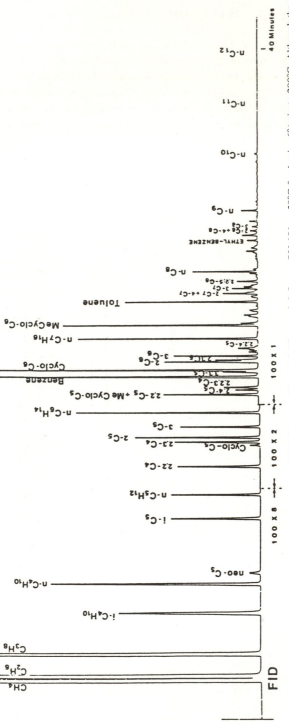

Fig. 9.21. Chromatogram of North Sea area natural gas [24]. Column, 50 m × 0.25 mm, d_f 0.5 μm, OV-101; −50°C for 1 min, 6°/min to 200°C. Although the goals of the two groups were different, variations that are clearly evident in the low-end and high-end separations as compared with those in Fig. 9.20 are probably attributable to differences in d_f, initial temperatures, and initial holds.

fluidlike diffusivities. Mooney *et al.* [23] have demonstrated the separation of low-molecular-weight hydrocarbons in a valved inlet system, using a relatively thick-film column at subambient conditions (Fig. 9.20). Thicker-film columns that have become available since that time have proved especially useful for the analysis of these low-boiling mixtures. Figure 9.21 illustrates a chromatogram of natural gas obtained from the North Sea area, chromatographed on a "standard" film capillary and programmed from −50 to 200°C [24]. In the refinery gas separation shown in Fig. 9.22, the very thick film combined with a relatively low

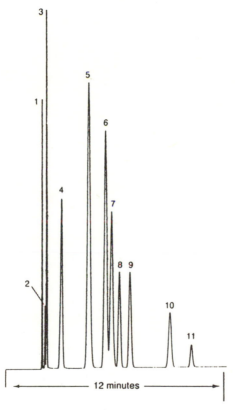

Fig. 9.22. Refinery gas on "very thick film" (d_f 5 μm) columns. Top, 30 m × 0.32 mm; bottom, 30 m × 0.53 mm; 0°C for 3 min, 10°/min to 50°C. Gas injection, 1 ml direct on-column with no special provision for cold trapping or band focusing; the lower pressure drop of the larger-diameter column permits the relatively massive injection to deposit shorter solute bands at the beginning of the column, partially compensating for the higher values of *h*. Solutes: (1) methane, (2) ethylene, (3) ethane, (4) propane, (5) isobutane, (6) 1-butene, (7) *n*-butane, (8) *trans*-2-butene, (9) *cis*-2-butene, (10) isopentane, (11) *n*-pentane. (Reprinted with permission of the copyright holder, J&W Scientific, Inc.)

starting temperature shifted the solute partition ratios enough to achieve the resolution of methane, ethylene, and ethane at 0°C.

Interesting separations have been shown on porous layer open tubular (PLOT) columns coated with alumina (usually as Al_2O_3/KCl) Figure 9.23 shows chromatograms generated by low-molecular-weight hydrocarbons on a fused silica PLOT column of this type; the increased retentions permit separation of these low-molecular-weight solutes at appreciably higher temperatures, obviating the need for subambient operation. Ingenious multicolumn valved inlet systems,

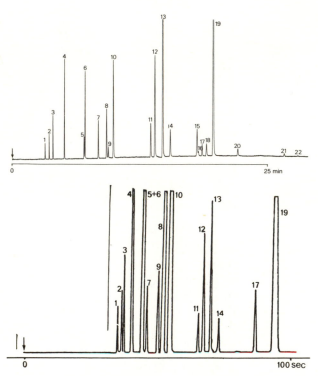

Fig. 9.23. Separation of C_1–C_5 hydrocarbons on a 50 m × 0.32 mm fused silica PLOT $Al_2O_3/$ KCl column. Top: nitrogen carrier at 1 atm. Temperature, 70–200°C at 3°/min. Solutes: (1) methane, (2) ethane, (3) ethene, (4) propane, (5) cyclopropane, (6) propene, (7) ethyne, (8) isobutane, (9) propadiene, (10) *n*-butane, (11) *trans*-2-butene, (12) 1-butene, (13) isobutene, (14) *cis*-2-butene, (15) isopentane, (16) 1,2-butadiene, (17) propyne, (18) *n*-pentane, (19) 1,3-butadiene, (20) 3-methyl-1-butene, (21) vinylacetylene, (22) ethylacetylene. Bottom: high-speed analysis of C_1–C_4 hydrocarbons. Column, 50 m × 0.32 mm fused silica PLOT Al_2O_3/KCl; hydrogen carrier at 3 atm; isothermal at 130°C. Solutes: (1) methane, (2) ethane, (3) ethene, (4) propane, (5) cyclopropane, (6) propene, (7) ethyne, (8) isobutane, (9) propadiene, (10) *n*-butane, (11) *trans*-2-butene, (12) 1-butene, (13) isobutene, (14) *cis*-2-butene, (15) isopentane, (16) 1,2-butadiene, (17) propyne, (18) *n*-pentane, (19) 1,3-butadiene. (Reproduced by permission of Chrompak USA.)

with potential for backflushing and/or redirecting individual fractions to open tubular, micropacked, and PLOT columns, have been designed (e.g., [24]) and are capable of achieving separations that are difficult if not impossible to duplicate with conventional equipment and/or conventional open tubular columns.

Figure 9.24 shows a chromatogram typical of a naphtha separation, and Fig. 9.25 illustrates gasoline feedstock patterns, replotted by an automated analytical (P.I.A.N.O.) system. Multidimensional chromatography has been used for the separation of some minor components often used as antiknock additives in unleaded gasoline (e.g., lower-molecular-weight alcohols, diisopropyl ether, methyl *tert*-butyl ether; Fig. 9.26 [25]). Levy and Yancey described an automated system for quantifying oxygenated additives on the basis of retention characteristics when gasoline injections were simultaneously split to two dissimilar columns (Fig. 9.27) [26].

Separations of weathered gasolines have significance for arson analysis, and methods of isolating samples from arson residues have been described [27]; chromatograms typical of such analyses are shown in Fig. 9.28.

The components of crude oil extend into the higher-molecular-weight range and require the use of less retentive columns (e.g., thinner film), capable of higher-temperature operation. Beyond generating a ''composite picture'' (Fig. 9.29 [27a], such separations are employed as ''fingerprints'' to establish the source of oil spills, in the dating of crudes by measurement of the C_{17}/pristane and C_{18}/phytane ratios (Fig. 9.30 [28]), and in characterizing steroidal hydrocar-

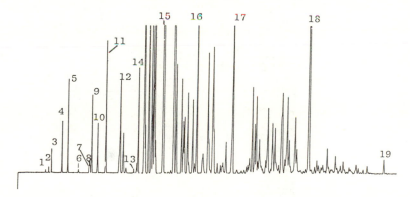

Fig. 9.24. Chromatogram produced by split injection of 1 μl of neat naphtha on a 30 m × 0.32 mm column with a 1-μm film of dimethyl silicone. Hydrogen carrier at 38 cm/sec (100°C); 28°C for 4 min, 2.5°/min to 100°C. Total analysis time 28 min. Attenuation: ×1 through *n*-butane, ×4 through 3-methylpentane; ×8 preceding benzene. Solutes: (1) propane, (2) 2-methylpropane, (3) *n*-butane, (4) 2-methylbutane, (5) *n*-pentane, (6) 2,2-dimethylbutane, (7) cyclopentane, (8) 2,3-dimethylbutane, (9) 2-methylpentane, (10) 3-methylpentane, (11) *n*-hexane, (12) methylcyclopentane, (13) benzene, (14) cyclohexane, (15) *n*-heptane, (16) toluene, (17) *n*-octane, (18) *n*-nonane, (19) *n*-decane. Reprinted with permission of the copyright holder, J&W Scientific, Inc.)

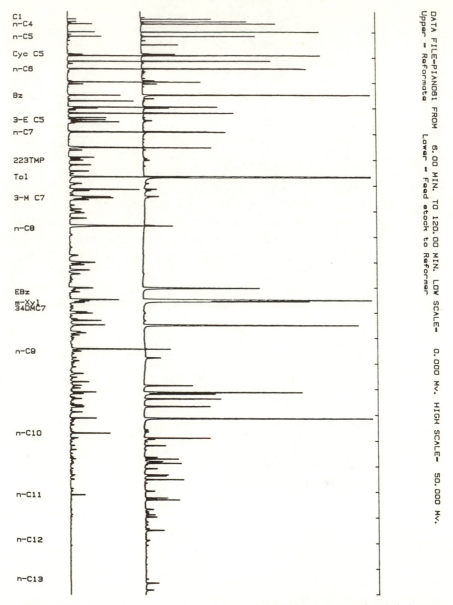

PIANO ANALYSIS: CHROMATOGRAMS
Replot Using 'Overlay' Mode of XY-PLOT

Fig. 9.25. Chromatograms of gasoline feedstocks, replotted by an automated analytical system (P.I.A.N.O.). (Courtesy of N. Johansen, Analytical Automation Specialists, Baton Rouge LA.)

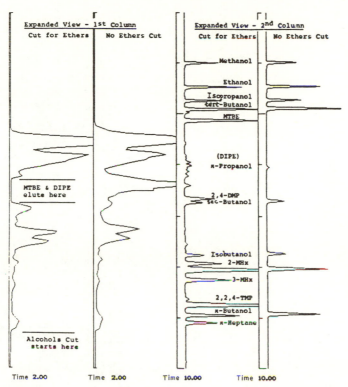

Fig. 9.26. Isolation of specific antiknock additives from unleaded gasoline by multidimensional gas chromatograph [25].

bons that are common constituents of crude oils and ancient sedimentary rocks [29]. The latter can be of geological interest; a sample displaying a high content of the C_{29} steranes (e.g., isomers of 24-ethylcholestane) as compared to the C_{27} steranes would suggest that the precursor organic matter contained land-derived biochemical compounds rather than those derived from marine organisms [30]. Capillary gas chromatography/mass spectrometry can also be employed to characterize shales and fossils isolated therefrom. Figure 9.31 shows a chromatogram of the extract of a fossil wood found in Posidonomia shale [31].

The upper operating temperature limit of the gas chromatographic column would seem to impose some restrictions on the analysis of crude oils and other samples that range into very high molecular weight (and low volatility) solutes. With the fused silica column, the protective outer polyimide coating is most temperature-limiting and begins to show signs of degradation at about 340°C; the rate of degradation increases rapidly as the temperature is increased. Conventional glass capillary columns, which do not require this protective sheath, are sometimes used for these very high temperature applications. The greater inci-

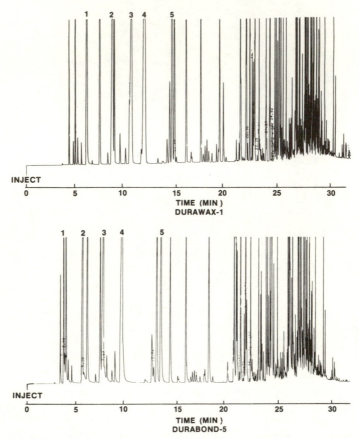

Fig. 9.27. Quantitation of oxygenates in unleaded gasoline by simultaneous split injection to two dissimilar columns [26]. Identified solutes: (1) methanol, (2) ethanol, (3) isopropanol, (4) *tert*-butanol, (5) methyl *tert*-butyl ether; levels between 0.8 and 1%.

dence of active sites on conventional glasses can be a problem with some solutes but is normally of scant significance in hydrocarbon analysis; column fragility remains a problem.

Unflawed fused silica tubing would be extremely resistant to breakage; however, some flaws persist and carry forward from the blank to the drawn tubing (Chapter 2), and others originate if the tubing is exposed to water vapor [32]. Under ideal conditions, metal coatings that are completely impervious to water can be applied to fused silica, and flaw-free fibers or tubing so protected would be highly resistant to mechanical breakage. Under conditions of temperature cycling, the difference in thermal expansion of the metal and silica could lead to complications. Polyimide coatings are highly resistant but not completely imper-

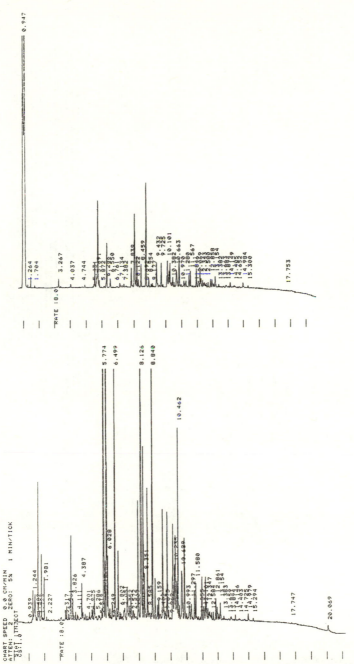

Fig. 9.28. Chromatograms produced by weathered gasoline using (top) "conventional" charcoal wire technique and (bottom) direct headspace sampling; both samples drawn in the presence of water vapor [27]. With dry sampling conditions, results from the two methods are comparable; response intensity of the charcoal wire technique is adversely affected by water vapor, while that of the headspace technique is not affected. This can be significant with samples that are more typically water-drenched.

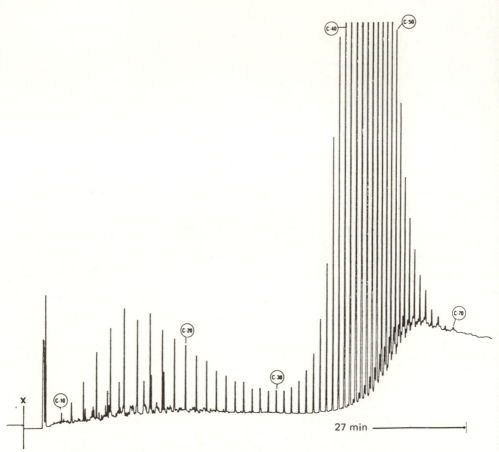

Fig. 9.29. Chromatogram produced by injection of a Texas crude oil on a 15 m × 0.53 mm aluminum-sheathed fused silica column coated with a 0.25-μm film of methyl silicone. Sample, 3 μl of a solution containing 20 mg crude/ml CS_2, programmed temperature vaporizing injection split 1 : 10; hydrogen carrier at 3 ml/min. Temperature profile, 60–400°C at 15°/min, hold. Attenuation ×64. Reprinted with permission from Lipsky and Duffy [272].

vious to water. However, it is probable that during its application, the polyimide (as contrasted with metal) is better able to flood and seal existing flaws that would otherwise grow at a stress-proportional rate [32] and eventually lead to breakage. Because commercial fused silica tubing does have surface flaws, the polyimide-coated tubing is generally stronger and more durable than the aluminum-coated tubing commonly available today [33].

In addition, the thermal stability of the polyimide coating is exceeded only slightly by that of the stationary phase, which imposes the next limitation on some (e.g., hydrocarbon) high-temperature analysis. This limitation is followed

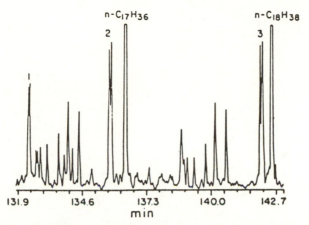

Fig. 9.30. Section of a chromatogram of a petroleum sample showing the separation of nor-pristane (1), pristane (2), and phytane (3) [28]. Column, 140 m × 0.27 mm, coated with $C_{87}H_{176}$ and programmed from 100 to 230°C at 0.8°/min; temperature range of the portion illustrated, 132–143°C.

rather closely by the upper temperature limit of the oven of the gas chromato-graph, only a few of which can operate at temperatures exceeding 375–400°C. The lability of most organic compounds at these higher temperatures is even more ominous; many chemical bonds (e.g., C–O, C–N, C–S) do not have the thermal stabilities necessary to survive exposure to these extreme temperatures. Hence the gas chromatographic separation may begin with well-defined solutes and end with their pyrolysis products. The quantitative and qualitative validities of such analyses are improved by employing chromatographic techniques not contingent on heat-induced volatilities of low-volatility thermally labile solutes (e.g., liquid chromatography, supercritical fluid chromatography).

Although aluminum-clad columns are not an analytical panacea, their utility for the analysis of higher-boiling thermally stable solutes (such as the higher-molecular-weight hydrocarbons at the upper end of crude oil samples) has been demonstrated. Using cold on-column injections, hydrocarbons as large as C_{87} and polywaxes up to C_{120} have been chromatographed. Degradation was not observed with either of these solutes, but the absence of degradation could be verified only on the hydrocarbon [34].

Lee and co-workers explored the suitability of some specially synthesized stationary phases for the separation of polycyclic compounds isolated from coal tars (e.g., [35–37]. Figure 9.32 shows chromatograms generated by coal tar fractions on two such stationary phases [38].

In spite of the considerable progress that has been made in column deactiva-tion, difficulties with extremely active solutes are not uncommon. Because thick-er stationary phase films tend to mask active sites on the tubing surface, the very

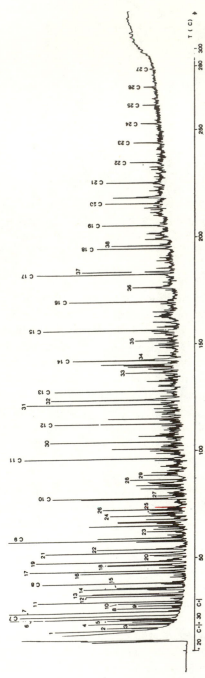

Fig. 9.31. Extract of a fossilized wood sample on a 30 m × 0.25 mm dimethyl polysiloxane column [31]. Temperature, 20°C for 3 min, 30°C for 3 min, 4°/min to 300°C, hold. Paraffinic hydrocarbons indicated by "C" prefix; other indicated solutes: (1) methylcyclopentane, (2) cyclohexane, (3) benzene, (4) 2-methylhexane, (5,6) dimethylcyclopentane isomers, (7) methylcyclohexane, (8) ethylcyclopentane, (9,10) trimethylcyclopentane isomers, (11) toluene, (12) 2-methylheptane, (13) dimethylcyclohexane, (14) methylethylcyclopentane, (15) dimethylcyclohexane, (16) ethylcyclohexane, (17) trimethylcyclohexane, (18) ethylbenzene, (19) m- and p-xylenes, (20) alkyl-substituted cyclohexane, (21) o-xylene, (22) methylethylcyclohexane, (23) propylcyclohexane, (24) propylbenzene, (25) methylethylbenzene, (26) butylbenzene, (27,28) alkylbenzenes, (29) naphthalene, (30) tetramethylbenzene, (31) 2-methylnaphthalene, (32) 1-methylnaphthalene, (33,34) dimethylnaphthalene isomers, (35) C$_{16}$ isoprenoid, (36) C$_{18}$ isoprenoid, (37) 2,6,10,14-tetramethylpentadecane (pristane), (38) 2,6,10,14-tetramethylhexadecane (phytane). (Reprinted with permission from *Chromatographia*.)

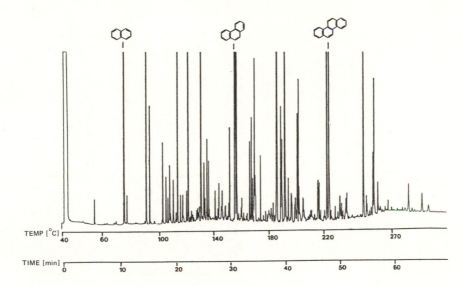

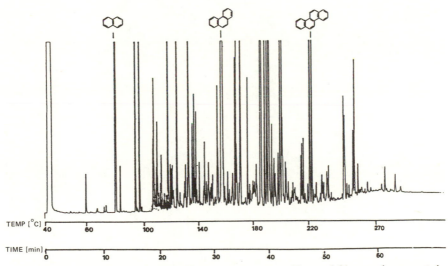

Fig. 9.32. Chromatograms generated by coal tar extracts on 12 m × 0.31 mm columns coated with (top) 25% biphenyl and (bottom) 50% phenyl polysiloxane stationary phases [38]. Hydrogen carrier at 50 cm/sec; 40–270°C at 4°/min.

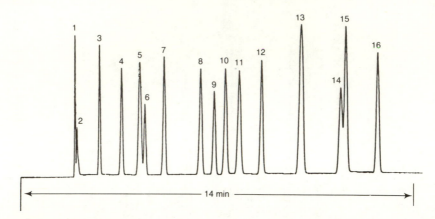

Fig. 9.33. Short-chain alcohols on a 30 m × 0.32 mm column with a 5.0-μm dimethyl silicone film. Split injection; 30°C for 0.5 min, 5°/min to 110°C. Solutes: (1) methanol, (2) acetaldehyde, (3) ethanol, (4) 2-propanol, (5) 2-methyl-2-propanol, (6) methyl acetate, (7) 1-propanol, (8) 2-butanol, (9) ethyl acetate, (10) 2-methyl-1-propanol, (11) 2-methyl-2-butanol, (12) 1-butanol, (13) 2-pentanol + 3-pentanol, (14) 3-methyl-1-butanol, (15) 2-methyl-1-butanol, (16) 1-pentanol. (Reprinted with permission of the copyright holder, J&W Scientific, Inc.)

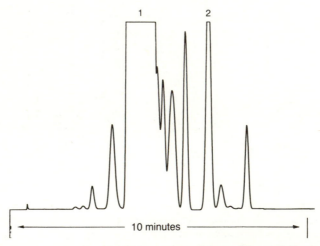

Fig. 9.34. Solvent impurity analysis: 99.7% benzene. Column, 30 m × 0.53 mm with a 5-μm dimethyl polysiloxane film. Helium at 20 ml/min, 1 μl direct injection, 45°C for 8 min, 5°/min to 100°C. Peak 1 is benzene; peak 2 is toluene. (Reprinted with permission of the copyright holder, J&W Scientific, Inc.)

thick film columns are usually even more inert than fused silica capillaries of "conventional" film thickness and can be used to advantage in the analysis of some active solutes, provided those solutes are sufficiently low boiling that the resulting increase in k does not hinder their elution. As an example, Fig. 9.33 shows a series of short-chain alcohols on a thick-film column.

Solvent assay can be a challenging problem, because a massive amount of the principal constituent must be introduced into the column if the detector is to respond to the minor constituents; i.e., the dynamic range of the system (and especially of the column) must be sufficiently large to accommodate this great quantitative difference in the components. Figure 9.34 shows an analysis of "99.7% pure benzene" on a 0.53 mm i.d. column coated with a 5-μm film of stationary phase. The low end of the dynamic range of a system is dictated by the level of detector "noise," while the upper end of the dynamic range is limited by column overload. In this case, the increased amount of stationary phase (contributed by both the larger column diameter and the thicker stationary phase film) has greatly increased the "sample accepting capacity" of the column without affecting the detector sensitivity. Hence the dynamic range of the system has been greatly expanded. The lower phase ratio of this column ($r/2d_f = 265$ μm/10 μm $= 26.5$) forced an increase in solute partition ratios, which was beneficial to the separation of these low-k compounds but would be a severe handicap in the analysis of impurities in higher-boiling solvents.

9.4 Environmental Applications

The environmental analyst may be concerned with the detection and/or quantitation of many different substances in a diversity of matrices. These may range from pesticide and herbicide chemicals, to polynuclear aromatic hydrocarbons (PAHs), to chlorinated compounds from vinyl chloride to polychlorinated biphenyls, in air, water, and soils. In many cases, the initial problem is one of sample preparation. Water samples, for example, may range from sources of drinking water to industrial wastewater to seepage from chemical dumps and disposal sites.

In many cases, regulatory agencies have specified a "standard" or "accepted" procedure for the analysis of given materials in a given matrix; with the large number of creative scientists engaged in these analytical activities, it is not surprising to find that our capabilities have progressed rapidly. With few exceptions, the "official methods of analysis" are less accurate, less sensitive, and more time-consuming than methods that were developed since their adoption. Regulations of this type change only slowly, but they do change, and the alternative (usually open tubular column) technology is often utilized in the meantime as a confirming technique.

The U.S. Environmental Protection Agency (EPA) first specified procedures for monitoring industrial effluents in 1977 [39]; some of the approved methods were amended in 1980 [40]. Methods 601 through 613 separate into 13 classes the 113 organic pollutants to be monitored and specify (1) a gas chromatographic or liquid chromatographic column and conditions resolving the pollutants in that class and (2) the system to be used for their detection. Methods 601, 602, 603, and 624 are concerned with volatile pollutants and specify purge and trap procedures. Recent work indicate that many of these pollutants can be detected at the parts-per-billion and parts-per-trillion levels by direct injections of water or water headspace [41–43]. EPA Method 624 is a GC/MS method that specifies a column packed with 1% SP-1000 on Carbopak B. Methods for establishing the composition of a stationary phase tailored to optimize the separation of a given mixture of compounds, discussed in Section 8.5, permitted designing a new stationary phase, DB-1301, precisely tailored to optimize the relative retentions of those solutes [44,45]. The DB-624 column is coated with that phase, with both α values and the phase ratio optimized to yield the best possible separation of these solutes [21]. Figures 9.35 and 9.36 show improved chromatographic separations of the solutes specified in Method 624 at higher sensitivities and in shorter analysis times than are possible with the official packed column [21]. Figure 9.37 shows a separation of the purgeable aromatics of EPA Method 602 on a 30 m × 0.53 mm Megabore open tubular column coated with the DB-624 stationary phase [21]. All solutes are baseline resolved in an analysis time of

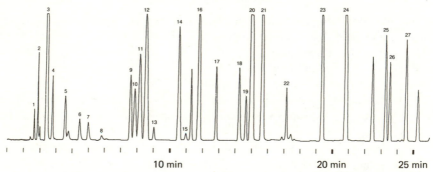

Fig. 9.35. Purgeable halocarbons (in methanol). Direct flash injection to a 30 m × 0.53 mm column with a 2-μm film of DB-1301 (DB-624 column) [21]; helium carrier at 5 ml/min; ECD; 35°C for 5 min, 5°/min to 140°C. Solutes: (1) chloromethane, (2) vinyl chloride, (3) bromomethane, (4) chloroethane, (5) 1,1-dichloroethylene, (6) methylene chloride, (7) trans-1,2-dichloroethylene, (8) 1,1-dichloroethane, (9) bromochloromethane, (10) chloroform, (11) 1,1,1-trichloroethane, (12) carbon tetrachloride, (13) 1,2-dichloroethane, (14) trichloroethylene, (15) 1,2-dichloropropane, (16) bromodichloromethane, (17) trans-1,3-dichloropropene, (18) cis-1,3-dichloropropene, (19) 1,1,2-trichloroethane, (20) tetrachloroethylene, (21) dibromochloromethane, (22) chlorobenzene, (23) bromoform, (24) 1,1,2,2-tetrachloroethane, (25) 1,3-dichlorobenzene (meta), (26) 1,4-dichlorobenzene (para), (27) 1,2-dichlorobenzene (ortho). (Reprinted from *American Laboratory*, volume 18, number 5, 1986, p. 60. Copyright 1986 by International Scientific Communications, Inc.)

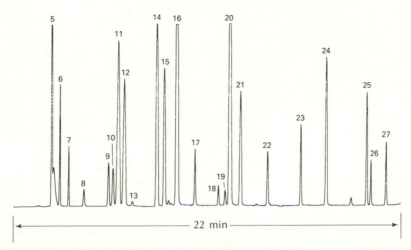

Fig. 9.36. Gas chromatographic analysis of the solutes specified in EPA Method 601, using a purge and trap sampler [21]. Conditions as in Fig. 9.35 except for the final temperature, 120°C, and column flow (and purge and trap flow), 8 ml/min helium. Same solutes as in Fig. 9.35, minus the first four. (Reprinted from *American Laboratory*, volume 18, number 5, 1986, p. 60. Copyright 1986 by International Scientific Communications, Inc.)

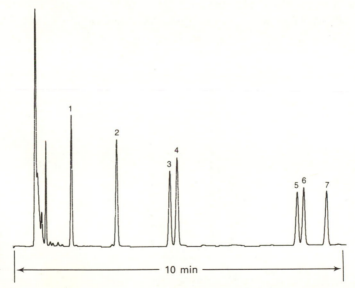

Fig. 9.37. Purge and trap gas chromatographic analysis of the solutes of EPA Method 602, purgeable aromatics (10 ppb) [21]. Same conditions as in Fig. 9.35 except for the column flow (and purge and trap desorption flow), 20 ml/min; temperature, 50°C for 1 min, 5°/min to 100°C; ECD. Solutes: (1) benzene, (2) toluene, (3) chlorobenzene, (4) ethylbenzene, (5) 1,3-dichlorobenzene, (6) 1,4-dichlorobenzene, (7) 1,2-dichlorobenzene. (Reprinted from *American Laboratory*, volume 18, number 5, 1986, p. 60. Copyright 1986 by International Scientific Communications, Inc.)

10.8 min; published chromatograms indicate an analysis time of 22 min with the approved packed column [46].

In the analysis of nitroaromatics and isophorone (EPA Method 609), two separate determinations are generally used on the official packed column; at 85°C, nitrobenzene and isophorone have been presented as severely tailing peaks on the solvent tail, isophorone emerging at about 5 min; at 145°C, the 2,6- and 2,4-dinitrotoluenes are well resolved with an analysis time of about 9 min [46]. All four solutes are resolved in a single pass in the open tubular separation shown in Fig. 9.38, with a total analysis time of 4.5 min.

Figure 9.39 shows separation of the phenols of EPA Method 625 on a 30 m × 0.53 mm (Megabore) open tubular column coated with a cross-linked, surface-bonded film of 5% phenyl, 1% vinyl polysiloxane; all solutes are baseline resolved in less than 11 min. Published separations with the approved packed column required just over 24 min and failed to resolve completely the 2,4-dinitrophenol from 4,6-dinitro-*o*-cresol [46].

The analytical advantages of the open tubular column have also been recognized by EPA scientists, some of whom pointed out that, as compared to packed column analyses, the capillary methods they investigated yielded good precision, excellent accuracy, and shortened analysis times, besides overcoming problems associated with sample complexity, large ranges of component concentrations, and the analysis of higher-boiling components [47].

Figure 9.40 shows the separation of a PAH standard. The DB-5 stationary

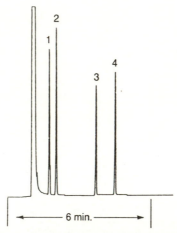

Fig. 9.38. Gas chromatographic analysis of nitroaromatics of EPA Method 609 on a 30 m × 0.53 mm column with a 1-μm film of trifluoropropyl methyl silicone; helium carrier at 10 ml/min; 165–220°C at 15°/min; FID. Solutes: (1) nitrobenzene, (2) isophorone, (3) 2,6-dinitrotoluene, (4) 2,4-dinitrotoluene. (Reprinted with permission of the copyright holder, J&W Scientific, Inc.)

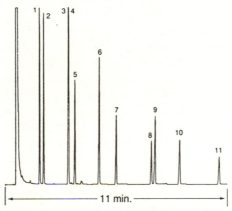

Fig. 9.39. Separation of the phenols specified in EPA Method 625 on a 30 m × 0.53 mm column with a 1.5-μm film of phenylmethyl polysiloxane; helium at 10 ml/min; 100–210°C at 10°/min; FID. Solutes: (1) phenol, (2) 2-chlorophenol, (3) 2-nitrophenol, (4) 2,4-dimethylphenol, (5) 2,4-dichlorophenol, (6) *p*-chloro-*m*-cresol, (7) 2,4,6-trichlorophenol, (8) 2,4-dinitrophenol, (9) 4-nitrophenol, (10) 2,6-dinitro-*o*-cresol, (11) pentachlorophenol. (Reprinted with permission of the copyright holder, J&W Scientific, Inc.)

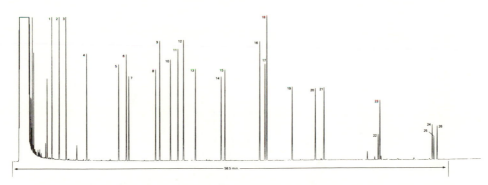

Fig. 9.40. Polycyclic aromatic hydrocarbon (PAH) standard (PAH 10). Splitless injection of 2 μl on a 30 m × 0.32 mm column with a 0.25-μm film of phenylmethyl polysiloxane; 40°C for 3 min, 4°/min to 320°C. Solutes: (1) *o*-xylene, (2) isopropylbenzene, (3) *n*-propylbenzene, (4) indane, (5) trimethylbenzene, (6) naphthalene, (7) benzothiophene, (8) 2-methylnaphthalene, (9) 1-methylnaphthalene, (10) triisopropylbenzene, (11) biphenyl, (12) dimethylnaphthalene, (13) hexamethylbenzene, (14) trimethylnaphthalene, (15) fluorene, (16) dibenzothiophene, (17) phenanthrene, (18) anthracene, (19) 1-methylphenanthrene, (20) fluoranthene, (21) pyrene, (22) benz[*a*]anthracene, (23) chrysene, (24) benzo[*e*]pyrene, (25) benzo[*a*]pyrene, (26) perylene. (Reprinted with permission of the copyright holder, J&W Scientific, Inc.)

phase, "equivalents" of which are listed in Table 9.1, contains just enough aromatic phenyl substitution to enhance the separation of these solutes. The same column is used to separate a series of pesticides and chlorinated aromatics in Fig. 9.41. Figure 9.42 shows the separation of another series of pesticides on a column under development to optimize the separation of solutes specified in EPA Method 608 [48]. The chromatogram of a sample derived from marine dredging, indicative of contamination and illustrating how the complexity of a "real-world" sample can interfere with separations, is shown in Fig. 9.43. By using a low-dead-volume tee to split the column effluent to differential detectors, additional information can be gained (Fig. 9.44 [49]). Peak identifications in PAH analyses are frequently based on retention characteristics and elution order; Huynh and Vu Duc [50] demonstrated that such assignments can be hazardous; changes in any number of parameters can affect elution orders of any non-homologous series separated under conditions of temperature programming (see Section 5.12). Schulze *et al.* [51] used capillary GC/MS to assign identifications to a number of oxygenated PAHs isolated from particulate matter emitted by automotive diesel. Liquid carbon dioxide was used as an extractant to minimize the loss of reactive compounds in determinations of substituted PAHs (NO_2-PAH, carbazoles, keto-PAHs, and aza-arenes) in aerosol samples by Stray *et al.* [52]. They also reported that capillary GC/MS (with negative-ion chemical

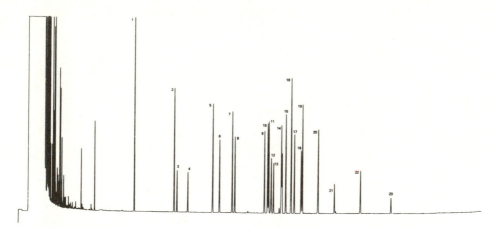

Fig. 9.41. Pesticides and chlorinated aromatics on a 30 m × 0.25 mm column, 0.25-μm film of 5% phenyl polymethylsiloxane. Splitless injection of 2 μl, ~10 ng/μl; FID. Temperature, 50°C for 2 min, 20°/min to 140°C, 4°/min to 300°C. Solutes: (1) technazine, (2) Cl_2-biphenyl, (3) hexachlorobenzene, (4) lindane, (5) Cl_3-biphenyl, (6) heptachlor, (7) Cl_4-biphenyl, (8) aldrin, (9) Cl_6-biphenyl, (10) *o,p*-DDE, (11) Cl_5-biphenyl, (12) α-chlordane, (13) *trans*-nonachlor, (14) dieldrin + *p,p'*-DDE, (15) *o,p*-DDD, (16) endrin, (17) *m,p*-DDD, (18) *p,p'*-DDD, (19) *o,p*-DDT, (20) *p,p'*-DDT, (21) Cl_7-biphenyl, (22) mirex, (23) Cl_8-biphenyl. (Reprinted with permission of the copyright holder, J&W Scientific, Inc.)

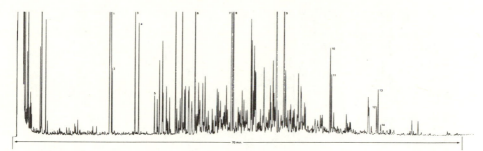

Fig. 9.42. Chromatogram of a chloroform extract of soil dredged from a marine waste disposal; column and chromatographic conditions as in Fig. 9.40. Solutes: (1) naphthalene, (2) benzothiophene, (3) 2-methylnaphthalene, (4) 1-methylnaphthalene, (5) biphenyl, (6) fluorene, (7) phenanthrene, (8) anthracene, (9) pyrene, (10) benz[a]anthracene, (11) chrysene, (12) benzo[e]pyrene, (13) benzo[a]pyrene, (14) perylene. (Reprinted with permission of the copyright holder, J&W Scientific, Inc.)

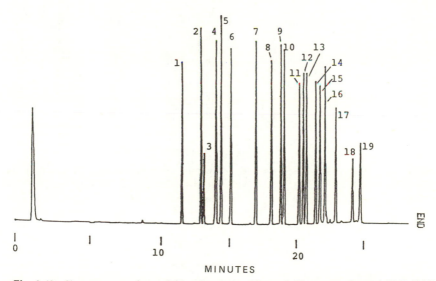

Fig. 9.43. Chromatogram of a pesticide mixture on a 30 m × 0.53 mm experimental "DB-608" column [48]. Helium carrier at 6 ml/min; 140°C for 0.5 min, 6°/min to 275°C; ECD with nitrogen makeup at 30 ml/min. Solutes (200 pg per component): (1) α-BHC, (2) β-BHC, (3) γ-BHC, (4) heptachlor, (5) δ-BHC, (6) aldrin, (7) heptachlor epoxide, (8) endosulfan-I, (9) 4,4'-DDE, (10) dieldrin, (11) endrin, (12) 4,4'-DDD, (13) endosulfan-II, (14) 4,4'-DDT, (15) endrin aldehyde, (16) endosulfan sulfate, (17) dibutyl chlorendate, (18) 4,4'-methoxychlor, (19) hexabromobenzene (surrogate). (Reprinted with permission of the copyright holder, J&W Scientific, Inc.)

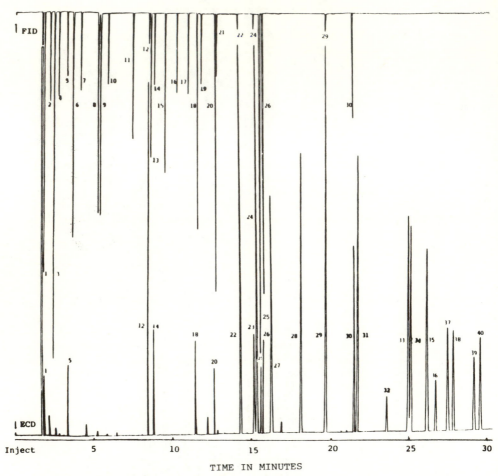

Fig. 9.44. Dual chromatograms generated by splitting the effluent from a 30 m × 0.25 mm column coated with 0.25 μm 5% phenyl polysiloxane to two detectors [49]; 40°C for 3 min, 15°/min to 90°C, 30°/min to 175°C (5 min hold), 3°/min to 250°C. Top, FID trace; bottom, ECD trace. Solutes: (1) acrylonitrile, (2) hexane, (3) benzene, (4) *n*-heptane, (5) *N*-nitrosodimethylamine, (6) toluene, (7) *n*-octane, (8) ethylbenzene, (9) *p*-xylene, (10) *n*-nonane, (11) *n*-decane, (12) nitrobenzene, (13) *n*-undecane, (14) isophorone, (15) *n*-dodecane, (16) *n*-tridecane, (17) *n*-tetradecane, (18) acenaphthalene, (19) *n*-pentadecane, (20) fluorene, (21) *n*-hexadecane, (22) α-benzene hexachloride, (23) β-benzene hexachloride, (24) γ-benzene hexachloride, (25) phenanthrene, (26) anthracene, (27) δ-benzene hexachloride, (28) heptachlor, (29) aldrin, (30) fluoranthene, (31) heptachlor epoxide, (32) α-endosulfan, (33) 4,4'-DDE, (34) dieldrin, (35) endrin, (36) β-endosulfan, (37) 4,4'-DDD, (38) endrin aldehyde, (39) endosulfan sulfate, (40) 4,4'-DDT.

ionization and $N_2/O/CH_4$ reaction gas mixtures) improved selectivity for the tetrachlorobenzo-*p*-dioxins versus pesticides and polychlorinated biphenyls (see below).

Polychlorinated biphenyls (PCBs), produced by chlorinating biphenyl to a specified weight percent chlorine, are obviously mixtures containing a number of positional isomers. The most commonly used PCBs were the Aroclors; with the exception of Aroclor 1016 (a PCB mixture containing 41% chlorine), the final

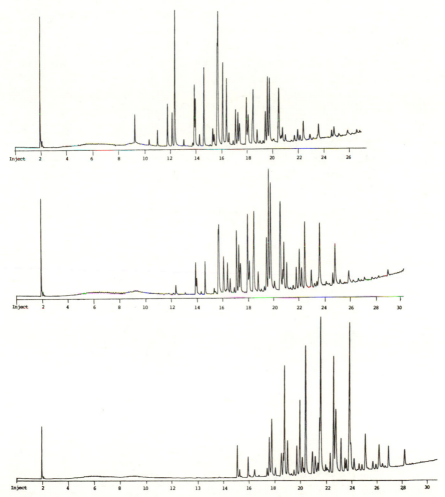

Fig. 9.45. Chromatograms of polychlorinated biphenyls (PCBs) on a 30 m × 0.25 mm column, 0.25-μm film of 5% phenylmethyl polysiloxane; 80°C for 2 min, 20°/min to 150°C, 4°/min to 275°C; ECD. Top, Aroclor 1232; center, Aroclor 1248; bottom, Aroclor 1254.

two digits specify the percentage (by weight) of chlorine. Because they were relatively stable compounds characterized by a high dielectric potential, PCBs were widely used for purposes as diverse as transformer and hydraulic fluids, and their occurrence in other formulations and oily waste products is not uncommon. Some authorities now argue that these compounds can be carcinogenic; other authorities view at least some of those claims as ill-founded. Nevertheless, the growing concern led to the development of techniques for monitoring uncharacterized fluids and spills (e.g., [53–55]). Typical Aroclor chromatograms are shown in Fig. 9.45. The PCBs also give rise to the tetrachlorodibenzo-*p*-dioxins (TCDDs); again, many isomers are possible. There are reports that the 2,3,7,8-isomer is both mutagenic and carcinogenic, but such claims are understandably difficult to authenticate with human subjects. Again, analytical techniques have developed; suspect samples are first screened by EPA Method 625 (which also isolates other contaminants), and positive samples are then subjected to EPA Method 613. Safe handling procedures, important for all toxic substances and of critical importance for 2,3,7,8-TCDD, are also discussed in the latter document. The gas chromatographic separation of the 2,3,7,8- isomer is not trivial, and Method 613 uses mass spectrometric detection for its differentiation. The gas chromatographic separation requires a polar stationary phase, a large number of theoretical plates, and relatively high temperatures. One such separation appears in Fig. 9.46.

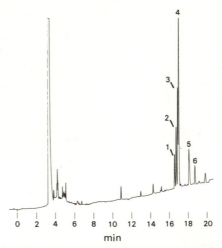

Fig. 9.46. Splitless injection of the critical tetrachlorodibenzodioxin isomers on a 60 m × 0.32 mm SP-2330 column; 200°C for 1 min, 3°/min to 250°C; helium carrier at 30 cm/sec (200°C); ECD. Numbered solutes (~0.5 ng per component): (1) 1,4,7,8-TCDD, (2) 2,3,7,8-TCDD, (3) 1,2,3,4-TCDD, (4) 1,2,3,7-TCDD plus 1,2,3,8-TCDD, (5) 1,2,7,8-TCDD, (6) 1,2,6,7-TCDD. (Reprinted with permission of Supelco, Inc., Bellefonte, Pennsylvania 16823, from the Supelco Reporter, volume IV, no. 4, July 1985, p. 1, Fig. A.)

9.5 Biological and Medical Applications

The low-level detection of compounds significant in this field poses stringent requirements for well-deactivated analytical systems; this is, of course, also true for some other active solutes, such as pesticides. These high levels of solute activity may necessitate not only the selection of well-deactivated columns but also attention to other parts of the analytical system; in particular, user deactivation of the sample injection area may be desirable (see Chapter 3). Where quantitation is important, techniques explored by Schomburg [56] and by Dandeneau and Zerrener [57] should be employed to establish that the reliability of the system extends to the levels measured for the solute of concern. For thermally labile solutes, cold on-column injection may also be required.

Recommendations for suitable stationary phases gleaned from suppliers' catalogs for samples fitting this class could be consolidated as in Table 9.3.

It is often desirable to monitor the level of an anesthetic, both in the air (or oxygen) stream provided to a patient and sometimes (as a safeguard against leakage) in the room atmosphere. Rapid analysis is essential, and thermal conductivity is the most practical means of detection. Figure 9.47 illustrates the use of a very thick film large-diameter open tubular column to achieve the separation of several anesthetics in less than 7 min; the packed column analysis that this was designed to replace required 28 min for this separation.

Poisoning by volatile organic solvents is sometimes evidenced by the odor of the patient's clothing or breath, but in other cases the diagnosis is facilitated by

TABLE 9.3

Columns Commonly Used for Biological and Medical Applications

Solutes	Recommended packed column by class	Open tubular type
Clinical/biomedical		
Ketosteroids (TMS)[a]	Silar 5CP	225
Cholesterol, free steroids, estrogens (tent-butyl DCMS)[b]	OV-17	17
Blood alcohols	Carbopak/SP-1000	PEG
Drugs		
Amphetamines	Apiezon/KOH	
Antidepressants	SP-2250	
Barbiturates	SP-2250, OV-17/H_3PO_4	
Alkaloids	SP-2250	1, 5, 17
Antiepileptics	SP-2110, SP-2510	
Diuretics	OV-17/H_3PO_4	

[a]TMS, trimethylsilyl derivative.
[b]DCMS, dichloromethyl silane derivative.

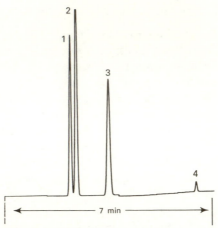

Fig. 9.47. Separation of an anesthetic mixture on a 30 m × 0.53 mm column, 5-μm film of dimethyl polysiloxane. Helium carrier at 6 ml/min; 50°C for 3.75 min, 30°/min to 125°C; TCD. Solutes: (1) isoflurane, (2) enflurane, (3) halothane, (4) methoxyflurane. (Reprinted with permission of the copyright holder, J&W Scientific, Inc.)

blood headspace analysis [58]. The more common agents of abuse, whose sources can include solvents (used per se or from sniffing glue, lacquer, etc.), aerosol propellents, or anesthetics, may include bromochlorodifluoromethane, n-butane, carbon tetrachloride, chlorobutanol, cryofluorane (Halon 114), ethyl acetate, halothane, isobutane, isopropanol, isopropyl nitrate, methyl ethyl ketone, propane, tetrachloroethylene, toluene, 1,1,1-trichloroethane, 2,2,2-trichloroethanol, trichloroethylene, and trichlorofluoromethane (Halon 11). Ramsey and Flanagan used headspace injections on two dissimilar packed columns and simultaneous flame ionization–electron capture detection (FID–ECD) to facilitate rapid detection and identification of these solutes in blood samples of acutely poisoned patients [58]. It is probable that the DB-624 column could accomplish such analyses even more rapidly. The separation of those solutes that typically occur in an analysis for blood alcohols is shown in Fig. 9.48; analyses of this type can also be performed by headspace sampling procedures, with great benefits to column lifetime.

Drugs are often among the more active solutes, and derivatization is normally required for their analysis on packed columns. The increased inertness of the fused silica open tubular column may allow direct analysis of the underivatized drug. In some cases (e.g., some of the antiepileptics), derivatization is employed to impart thermal stability and is required for both packed and open tubular columns. A separation of some tricyclic antidepressant drugs is shown in Fig. 9.49; the other drug analyses shown (Figs. 9.50–9.55) all utilized the same large-diameter open tubular column.

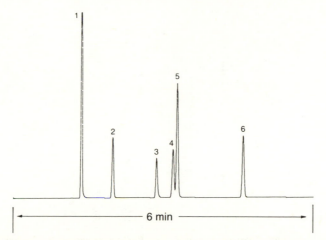

Fig. 9.48. Blood alcohols on a 30 m × 0.53 mm column, 1-μm bonded PEG; 0.2 μl injection of water containing 0.1% each solute. Helium carrier at 7 ml/min; 40°C for 2 min, 10°/min to 80°C. Solutes: (1) acetaldehyde, (2) acetone, (3) methanol, (4) isopropanol, (5) ethanol, (6) *n*-propanol. (Reprinted with permission of the copyright holder, J&W Scientific, Inc.)

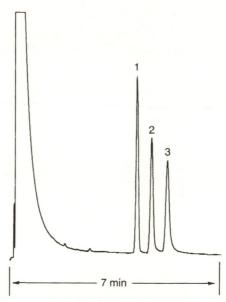

Fig. 9.49. Chromatogram of underivatized tricyclic antidepressant drugs on a 15 m × 0.53 mm column with a 1-μm film of 50% phenyl polysiloxane. Direct injection of 1 μl containing 300 ng of each drug in methanol. Helium carrier at 20 ml/min; 150–220°C at 20°/min; FID. Solutes: (1) amitriptyline, (2) nortriptyline, (3) protriptyline. (Reprinted with permission of the copyright holder, J&W Scientific, Inc.)

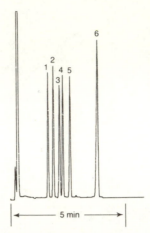

Fig. 9.50. Underivatized barbiturates on a 30 m × 0.53 mm column, 1.5-μ methyl polysiloxane. Direct injection of 1 μl containing 1 μg each in methanol. Helium carrier at 30 ml/min; 175–210°C at 8°/min. Solutes: (1) aprobarbital, (2) butabarbital, (3) amobarbital, (4) pentobarbital, (5) secobarbital, (6) phenobarbital. (Reprinted with permission of the copyright holder, J&W Scientific, Inc.)

In studies of the derivatization of difunctional amines (e.g., β-adrenergic blocking drugs), Jacob *et al.* established that with the exception of ephedrine, all drugs examined produced two stereoisomeric derivatives that were resolved to well-formed peaks under the conditions employed [59]. Alm *et al.* used simultaneous injection of drug mixtures onto two dissimilar columns, one employing FID and the other nitrogen/phosphorous detection (NPD), to generate chromatographic data of the type shown in Fig. 9.56 [60]. A computerized system of

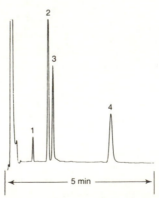

Fig. 9.51. Alkaloid drugs; conditions as in Fig. 9.50, except 250 ng each component, isothermal at 255°C. Solutes: (1) cocaine, (2) codeine, (3) morphine, (4) quinine. (Reprinted with permission of the copyright holder, J&W Scientific, Inc.)

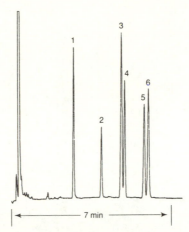

Fig. 9.52. Sedative and hypnotic drugs; conditions as in Fig. 9.50, except 70 ng per component; 150–250°C at 10°/min. Solutes: (1) aprobarbital, (2) meprobamate, (3) diphenhydramine, (4) mephobarbital, (5) methapyrilene, (6) chlorpheniramine. (Reprinted with permission of the copyright holder, J&W Scientific, Inc.)

identification, based on solute retentions relative to those of two internal standards, was also described. A number of underivatized barbiturate and alkaloid drugs, neat and in spiked urine, were separated by Plotczyk and Larson [61], who also assigned retention indices and studied the effects on column lifetime of injections of polar solutions of such solutes.

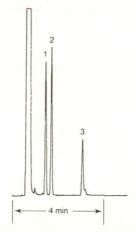

Fig. 9.53. Amphetamines; conditions as in Fig. 9.50, except 500 ng per component, helium at 14 ml/min, 150°C isothermal. Solutes: (1) amphetamine, (2) methamphetamine, (3) ephedrine. (Reprinted with permission of the copyright holder, J&W Scientific, Inc.)

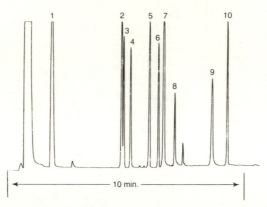

Fig. 9.54. Anticonvulsant drugs; conditions as in Fig. 9.50, except 200 ng each component, helium at 8 ml/min; 160°C for 2 min, 15°/min to 275°C. Solutes: (1) ethosuximide, (2) methsuximide, (3) phensuximide, (4) n-desmethylmethsuximide, (5) phenytoin, (6) 5-methyl-5-phenylhydantoin, (7) phenylethylmalonamide, (8) phenobarbital, (9) primidone, (10) carbamazephine. (Reprinted with permission of the copyright holder, J&W Scientific, Inc.)

Figure 9.57 shows results presented by Lutz *et al.* [62], who developed an automated procedure for the determination of isosorbide dinitrate and its metabolites in blood plasma, using capillary gas chromatography and ECD. A chromatogram of the total acid fraction of a urine sample from a healthy female is shown in Fig. 9.58, and Fig. 9.59 examines the less abundant components of the same sample following removal of the major components by liquid chromatography [63]. Chromatographic parameters yielding the separation of the trimethylsilyl (TMS) derivatives of several steroids were established by Knorr *et al.* [64]

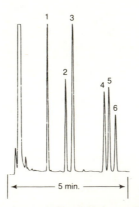

Fig. 9.55. Antihistamines; conditions as in Fig. 9.50, except 85 ng per component, helium at 18 ml/min, 215–275°C at 5°/min. Solutes: (1) pheniramine, (2) tripelennamine, (3) cyclizine, (4) pyrilamine, (5) triprolidine, (6) promethazine. (Reprinted with permission of the copyright holder, J&W Scientific, Inc.)

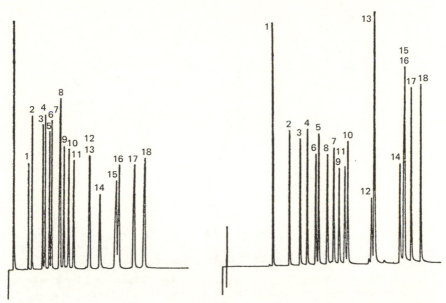

Fig. 9.56. Simultaneous dual-column, dual-detector analysis of an illicit drug injection [60]. Left, 11 m × 0.2 mm column with a 0.26-μm film of 5% phenyl polysiloxane; FID. Right, 10 m × 0.2 mm column, 0.2-μm film of trifluoropropyl silicone; NPD. Both columns: 75°C for 2 min, 10°/min to 280°C, hold. Solutes: (1) metarbital, (2) barbital, (3) allobarbital, (4) aprobarbital, (5) propylamphetamine (internal standard I), (6) butalbarbital, (7) amobarbital, (8) nealbarbital, (9) pentobarbital, (10) vinbarbital, (11) secobarbital, (12) brallobarbital, (13) hexobarbital, (14) trioctylamine (internal standard II), (15) phenobarbital, (16) cyclobarbital, (17) alphenal, (18) heptabarbital.

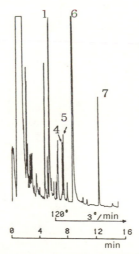

Fig. 9.57. Chromatogram generated from a human plasma extract after oral administration of isosorbide dinitrate [62]. Column, 30 m × 0.32 mm, 0.25-μm film of 5% phenyl polysiloxane; 60°C during injection, ballistic rise to 120°C, 3 min hold, 3°/min to 195°C, hold; hydrogen carrier; ECD. Solutes: (1) isosorbide-2-mononitrate, (4) isomannide mononitrate, (5) isoidide mononitrate, (6) isosorbide-5-mononitrate, (7) isosorbide dinitrate.

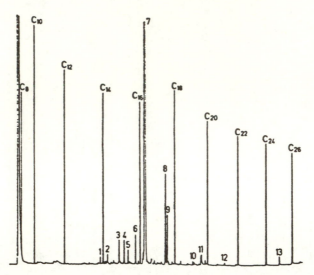

Fig. 9.58. Chromatogram of the total acid fraction of a urine sample from a healthy woman; sample extracted with chloroform and reacted with diazomethane. Column, 30 m × 0.33 mm coated with dimethyl silicone; hydrogen at 2 ml/min; FID; 80°C for 5 min, 2°/min to 280°C. Numbered solutes are mono-, di-, or trimethyl esters of the following acids: (1) 3-methyoxyphenylacetic, (2) citric, (3) 2,4-dimethyloctanedioic, (4) nonanedioic, (5) *N*-acetyl-2-aminooctanoic, (6) 3,4-dimethoxyphenylacetic, (7) hippuric, (8) indoleacetic, (9) *N*-methylhippuric, (10) palmitic, (11) phthalic, (12) stearic, (13) di(*n*-butyl) phthalate.

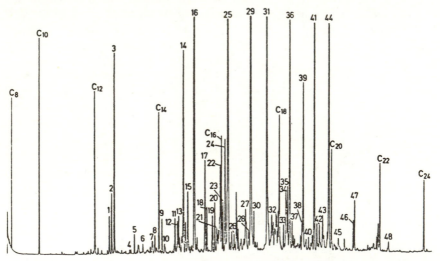

Fig. 9.59. Chromatogram of the minor constituents of Fig. 9.58 [63]; same conditions. Numbered solutes represent mono-, di-, or trimethyl esters of the following acids: (1) 2-methylhexanedioic, (2) 3-methylhexanedioic, (3) 2-hydroxyphenylacetic, (15) 3,5-dimethyloctanedioic, (16) nonanedioic, (17) 3,4-dimethyloctanedioic, (24) hexanedioic, (25) 5-decynedioic, (29) 3-methylfuran-2,5-dipropanoic, (31) 2-(3-carboxy-4-methyl-5-*n*-propylfuranyl)-*n*-propanoic, (41) 2-(3-carboxy-4-methyl-5-*n*-pentylfuranyl)-*n*-propanoic.

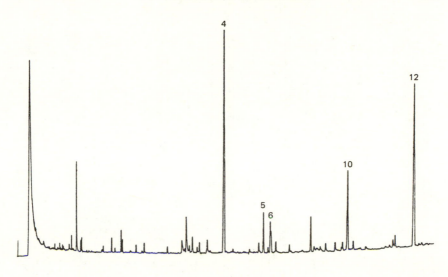

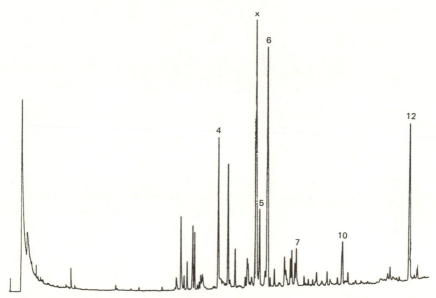

Fig. 9.60. Chromatograms of urinary steroids as trimethylsilyl ethers [64]; 35 m × 0.22 mm column coated with dimethyl silicone; 190–290°C at 2°/min; FID. Top, normal adult woman; solutes: (5) pregnanediol, (6) pregnanetriol, (10) tetrahydrocortisone. Bottom, insufficiently treated 12-year-old girl with 21-hydroxylase deficiency; solutes: (4) internal standard, (x) 17-hydroxypregnanolone, (6) pregnanetriol. (Reproduced with permission of S. Karger AG, Basel, from reference [64].)

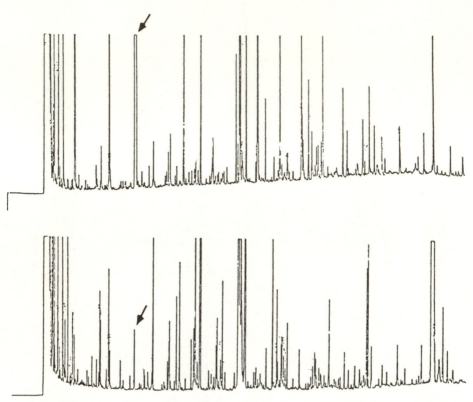

Fig. 9.61. Gas chromatographic profiles of saliva extracts [65] on a 5% phenyl polysiloxane column. Top, patient suffering liver disorder; bottom, normal patient. The diagnostic peak is indicated by the arrow.

and employed to demonstrate differences in the patterns of urinary steroids from normal and diseased patients (Fig. 9.60). Lochner *et al*. [65] suggested that a noninvasive technique, based on GC/MS patterns generated by saliva extracts, could be used in the detection of certain pathological states; patterns from saliva extracts of a normal and a diseased patient are shown in Fig. 9.61.

References

1. G. Takeoka, S. Ebeler, and W. Jennings, *ACS Symp. Ser.* **289,** 25 (1985).
2. E. Guenther, ''The Essential Oils,'' Vol. 3. Van Nostrand-Reinhold, Princeton, New Jersey, 1949.
3. J. W. Gramshaw and K. Sharpe, *J. Sci. Food Agric.* **31,** 93 (1980).
4. M. M. Bradley and J. W. Gramshaw, *J. Sci. Food Agric.* **31,** 99 (1980).
5. W. Jennings, *J. Chromatogr. Sci.* **22,** 129 (1984).
6. W. Mick and P. Schreier, *Agric. Food Chem.* **32,** 924 (1984).

7. T. Habu, R. A. Flath, T. R. Mon, and J. F. Morgan, *J. Agric. Food Chem.* **33,** 249 (1985).
8. K. H. Engle and R. Tressl, *J. Agric. Food Chem.* **31,** 998 (1983).
9. C. Macku, M.S. Degree in Agricultural Chemistry, University of California, Davis (to be published).
10. R. Wittkowski, Visiting Research Scientist at the University of California, Davis (to be published).
11. M. Guentert, Visiting Research Scientist at the University of California, Davis (to be published).
12. R. E. Kaiser and R. I. Rieder, *J. High Res. Chromatogr.* **8,** 863 (1985).
13. K. Markides, L. Blomberg, S. Hoffmann, J. Buijten, and T. Wannman, *J. Chromatogr.* **302,** 319 (1984).
14. E. Bayer, *Z. Naturforsch.* **38,** 1281 (1983).
15. G. Schomburg, I. Benecke, and G. Severin, *Proc. Int. Symp. Capillary Chromatogr., 6th, 1985,* p. 104 (1985).
16. C. E. Kientz and A. Verweij, *J. Chromatogr.* **355,** 229 (1986).
17. B. Kolb, P. Pospisil, and M. Auer, *Chromatographia* **19,** 113 (1984).
18. D. M. Wyatt, *J. Chromatogr. Sci.* **21,** 508 (1983).
19. G. Reineccius, personal communication (1985).
20. S. Jacobsson, *J. High Res. Chromatogr.* **7,** 185 (1984).
21. R. R. Freeman and R. M. A. Lautamo, *Am. Lab.* **18,** 60 (1986).
22. R. R. Freeman, personal communication (1986).
23. S. A. Mooney, F. P. Di Sanzo, and C. J. Lowther, *J. High Res. Chromatogr.* **5,** 684 (1982).
24. C. J. Cowper and P. A. Wallis, *Chromatographia* **19,** 85 (1985).
25. N. Johansen, *J. High Res. Chromatogr.* **7,** 487 (1984).
26. J. M. Levy and J. A. Yancey, *J. High Res. Chromatogr.* **9,** 383 (1986).
27. V. Reeve, J. Jeffery, D. Weihs, and W. Jennings, *J. Forensic Sci.* **31,** 479 (1986).
27a. S. R. Lipsky and M. L. Duffy, *HRC CC, J. High Res. Chromatogr. Chromatogor. Commun.* **9,** 376 (1986).
28. H. Borwitzky and G. Schomburg, *J. Chromatogr.* **240,** 307 (1982).
29. G. A. Warburton and J. E. Zumberge, *Anal. Chem.* **55,** 123 (1983).
30. B. P. Tissot and D. H. Welte, "Petroleum Formation and Occurrence." Springer-Verlag, Berlin and New York. 1978.
31. W. Bergmann, W. Heller, A. R. Hernanto, M. Schallies, and E. Bayer, *Chromatographia* **19,** 165 (1985).
32. T. A. Michalske and S. W. Freiman, *Nature (London)* **295,** 511 (1982).
33. G. Nelson (Polymicro Specialties), personal communication (1986).
34. G. Dupré (Exxon Research and Engineering, Arandale, New Jersey), personal communication (1986).
35. K. E. Markides, H.-C. K. Chang, C. M. Schrenberger, M. Nishioka, B. J. Tarbet, J. S. Bradshaw, and M. L. Lee, *Proc. Int. Symp. Capillary Chromatogr., 6th, 1985,* p. 137 (1985).
36. J. S. Bradshaw, S. J. Crowley, C. W. Harper, and M. L. Lee, *J. High Res. Chromatogr.* **7,** 89 (1984).
37. M. L. Lee, J. C. Kuei, N. W. Adams, B. J. Tarbet, M. Nishioka, B. A. Jones, and J. S. Bradshaw, *J. Chromatogr.* **302,** 303 (1984).
38. J. C. Kuei, J. I. Shelton, L. W. Castle, R. C. Kong, B. E. Richter, J. S. Bradshaw, and M. L. Lee, *J. High Res. Chromatogr.* **7,** 13 (1984).
39. *Fed. Regist.* **44,** No. 233 (1979).
40. *Fed. Regist.* **45,** No. 98 (1980).
41. K. Grob and A. Habich, *J. High Res. Chromatogr.* **6,** 11 (1983).
42. K. Grob, *L. Chromatogr.* **299,** 1 (1984).
43. M. F. Mehran, W. J. Cooper, M. Mehran, and W. Jennings, *J. Chromatogr. Sci.* **24,** 142 (1986).

44. M. F. Mehran, W. J. Cooper, and W. Jennings, *J. High Res. Chromatogr.* **7**, 215 (1984).
45. M. F. Mehran, W. J. Cooper, R. Lautamo, R. R. Freeman, and W. Jennings, *J. High Res. Chromatogr.* **8**, 715 (1985).
46. "Water Pollution Analyses and Standards: New EPA Procedures," Supelco GC Bull. 775B. Supelco Co., Inc. 1982.
47. F. A. Dreisch and T. O. Munson, *J. Chromatogr. Sci.* **21**, 111 (1983).
48. D. Kukla (J & W Scientific, Inc.) personal communication (1986).
49. M. F. Mehran, W. J. Cooper, M. Mehran, and R. Diaz, *J. High Res. Chromatogr.* **7**, 639 (1984).
50. C. K. Huynh and T. Vu Duc, *J. High Res. Chromatogr.* **7**, 270 (1984).
51. J. Schulze, A. Hartung, H. Kiess, J. Kraft, and K. H. Lies, *Chromatographia* **19**, 391 (1984).
52. H. Stray, S. Mano, A. Mikalsen, and M. Oehme, *J. High Res. Chromatogr.* **7**, 74 (1984).
53. "The Determination of PCBs in Transformer Fluid and Waste Oil," EPA 600/4-81-045. Environmental Monitoring and Support Laboratory, Office of Research and Development, USEPA, Cincinnati, Ohio, 1982.
54. "Methods for Organic Chemical Analysis of Municipal and Industrial Wastewater," EPA 600/4/82-057. Environmental Monitoring and Support Laboratory, Office of Research and Development, USEPA, Cincinnati, Ohio, 1982.
55. "Interim Methods for the Sampling and Analysis of Priority Pollutants in Sediments and Fish Tissue." Environmental Monitoring and Support Laboratory, Office of Research and Development, USEPA, Cincinnati, Ohio, 1980.
56. G. Schomburg, *J. High Res. Chromatogr.* **2**, 461 (1979).
57. R. D. Dandeneau and E. H. Zerrener, *J. High Res. Chromatogr.* **2**, 451 (1979).
58. J. D. Ramsey and R. J. Flanagan, *J. Chromatogr.* **240**, 423 (1982).
59. K. Jacob, G. Schnabl, and W. Vogt, *Chromatographia* **19**, 2161 (1985).
60. S. Alm, S. Jonson, H. Karlsson, and E. G. Sundholm, *J. Chromatogr.* **254**, 179 (1983).
61. L. L. Plotczyk and P. Larson, *J. Chromatogr.* **257**, 211 (1983).
62. D. Lutz, J. Rasper, W. Gielsdorf, J. A. Settlage, and H. Jaeger, *J. High Res. Chromatogr.* **7**, 58 (1984).
63. G. Spiteller, *Angew. Chem., Int. Ed. Engl.* **24**, 451 (1985).
64. D. Knorr, F. Bidlingmaier, and U. Kuhnle, *Horm. Res.* **16**, 201 (1982).
65. A. Lochner, S. Weisner, A. Zlatkis, and B. S. Middleditch, *J. Chromatogr.* **378**, 267 (1986).

CHAPTER 10
TROUBLESHOOTING

10.1 General Considerations

Elementary troubleshooting, as based on the shape and behavior of solute peaks, has been covered elsewhere [1,2]; the present chapter builds on a basic troubleshooting knowledge to considerations of other chromatographic problems and methods for their rectification.

New Column Installation

Most practitioners tend to condemn the column for defects such as an excessively rising or unsteady baseline, for ghost, malformed, or tailing peaks, and for many other chromatographic anomalies. These problems can also be due to extracolumn factors (e.g., deposits or residues in the inlet); active sites within the inlet can produce tailing peaks on even the best-deactivated column (see Chapter 7).

A new column that produces excellent results when first installed sometimes undergoes rapid deterioration, resulting in decreased resolution, tailing peaks, and/or a high rate of bleed. High-quality columns are both materiel- and labor-intensive; column manufacturing costs can be reduced by using lower-quality supplies (stationary phases, tubing), by devoting less time and effort to deactivating and coating the tubing and individually testing each finished column, and by lowering the minimum quality control performance standards so that a higher percentage of the finished columns can be shipped and sold. A given lot of columns produced in this manner will usually still include a few columns that are excellent, a number that are mediocre, and some of very poor quality. The manufacturing costs and initial purchase price of higher-quality columns are generally higher, but the columns can be expected to produce superior results for longer times when given proper care. The lifetime of a lower-quality column is usually preordained by manufacturing parameters and is only marginally affected by operational conditions. The lifetime of a high-grade column, on the other hand, is less limited by conditions of its manufacture and is very much influenced by the operational conditions; one of the more critical circumstances that can jeopardize column lifetime is the use of contaminated carrier gas.

Gas scrubbers and purifiers should be used on even high-purity gases; the presence of contaminants in the carrier gas reaching the column testifies to either expended gas scrubbers and filters or subsequent contamination of the clean gas by postfilter passage through dirty lines or soiled inlets. Prior to the installation of a new column, it is good general practice to recheck in-line gas scrubbers and filters, to clean the inlet liner and the injection block cavity in which it is housed, and to replace the septum. The column should require only brief conditioning (see Chapter 6), and the manufacturer's test chromatogram should be duplicated under the conditions specified. If the test chromatogram so generated is inferior to that enclosed with the column (in terms of component separation, peak symmetry and conformation, and/or bleed), it indicates inadeqacies in the equipment, in column installation, or in operational techniques. One or more of these factors is degrading column performance and may jeopardize column lifetime; particularly in the case of a high bleed, the cause should be ascertained and corrected before the column is subjected to routine use.

10.2   Use of Test Mixtures

Most test mixtures include solutes that embrace a range of functional groups; hydrocarbon solutes are almost always included. By evaluating the numbers of theoretical plates developed on hydrocarbon solutes and the shapes of those peaks, the analyst can establish the behavior of a nonactive solute under a particular set of conditions on that apparatus. Malformed hydrocarbon peaks testify to inadequacies that may include poor injection technique and gas flow problems (see also Section 10.7). The latter may be related to gross departures from the optimum column velocities or to poorly swept volumes in the gas flow path, i.e., excessive or dead volumes in the inlet or detector, an inadequate split ratio (split injection), flashback, insufficient makeup gas, or poor positioning of the column ends in the inlet and detector.

Well-formed hydrocarbon peaks serve as "standards" to which the peak shapes of active solutes can be compared. Figure 10.1 shows a typical column test mixture on three different columns; the top chromatogram, in which the peak shapes for hydrocarbon solutes are essentially the same as those for active solutes, is indicative of a well-deactivated column properly operated in good equip-

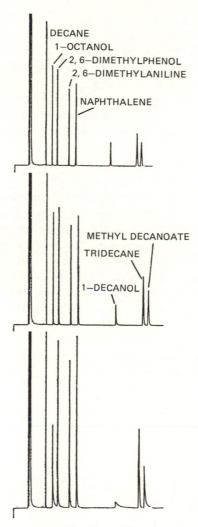

Fig. 10.1. Test chromatograms generated on three different new columns in the same instrument and under the same conditions. All columns 30 m × 0.32 mm, dimethyl polysiloxane; 105°C isothermal. Top, a well-deactivated column. Center, all nonhydrocarbon solutes exhibit some tailing; the decrease in the relative concentrations of the two alcohols indicates irreversible adsorption of this species. Bottom, excessive activity; note that hydrocarbon peaks are still reasonable.

ment. In the center chromatogram, both reversible and irreversible adsorption are evident on the two alcohol peaks. Reversible adsorption is evidenced as tailing, and the decrease in the areas of those peaks, relative to those of their hydrocarbon neighbors (compare to top chromatogram), testifies to irreversible adsorption and signals an alert against using this system for quantitation of these active solutes. In addition, peaks generated by the phenol, the aniline, and the ester are unsatisfactory. Because the same unit was used to generate the test chromatograms on all three columns, the activity can be blamed on the column rather than on the inlet or other parts of the chromatographic system. The bottom chromatogram shows a column with extreme activity where the peak shapes of active solutes, relative to those generated by hydrocarbons, are extremely poor. If this is a new column containing methyl silicone or methylphenyl silicone, it is possible that its performance may improve with resilylation (see below). Chromatograms of this type can also be generated by injection residues in the inlet or in the column; careful cleaning or removal of a few meters of column at the inlet end may resolve the problem (see below).

A chromatogram of a very demanding test mixture proposed by Grob *et al*. [3] is shown in Fig. 10.2. Under the standardized conditions (discussed in [4]), each peak should extend to reach a curved line drawn through the maxima of the nonsorbed solutes. In the example shown, abstraction of a portion of each of the solutes represented by peaks 3 (octanol), 4 (nonanal), 7 (2,6-dimethylaniline), and 10 (dicyclohexylamine) is obvious; the fronting evidenced by peak 6 is due not to activity but to a vapor phase overload (see below) and testifies to the unsuitability of the methyl silicone phase for carboxylic acid analysis (see also Section 6.2). The diminution of peaks 4 and 7 is probably due to aldehyde–aniline interaction producing the Schiff base in the test solution, a phenomenon that led to the recommendation that, where both solutes are desired, they should be presented in separate test mixtures [5].

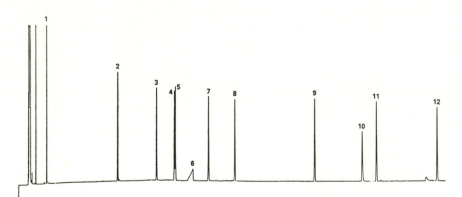

Fig. 10.2. Chromatogram of the Grob test mixture [3] on a 30 m × 0.25 mm column coated with 5% phenylmethylsiloxane; hydrogen carrier, t_M 60 sec at 40°C; 40–140°C at 1.67°/min. Solutes: (1) 2,3-butanediol, (2) *n*-decane, (3) 1-octanol, (4) nonanal, (5) 2,6-dimethylphenol, (6) 2-ethylhexanoic acid, (7) 2,6-dimethylaniline, (8) *n*-dodecane, (9) methyl decanoate, (10) dicyclohexyl amine, (11) methyl undecanoate, (12) methyl dodecanoate. See text for discussion. (Reprinted with permission of the copyright holder, J&W Scientific, Inc.)

10.3   Column Bleed

Column bleed is characterized by a rising baseline as the column temperature is increased and may in some cases be detector-specific; aspiration of a small amount of phosphate-containing leak detection fluid into a column can cause that column to produce baselines that are acceptable by flame ionization detection (FID) but wildly erratic with nitrogen/phosphorus detection (NPD); contamination of the detector itself at the column attachment fitting can also lead to high noise levels and/or temperature-related signal that can be confused with column bleed.

Figure 10.3 shows a chromatogram with a serious bleed problem. This relatively common phenomenon was formerly attributed to the distillation of stationary phase from the column. Today we recognize that while bleed attests to material entering the detector, there are several possible origins of that material and several plausible explanations for its presence in or generation from some of those sources.

Isolating Bleed Problems

When a bleed problem becomes evident, the column should be removed from the oven and the column attachment fittings at both the inlet and the detector should be closed with approved solid plug fittings; the detector should then be activated, the detector signal trace energized, and the oven temperature increased to the level where problems were previously experienced.* An increase in baseline indicates that the signal is extracolumn and probably results from a dirty detector or residues condensed in the lines supplying makeup and/or combustion hydrogen to the detector. The detector should be disassembled and cleaned according to the manufacturer's instructions, and all lines, in-line valves, and flow controllers supplying the detector should be carefully cleaned; in-line gas scrubbers should be replaced. Finally, the efficacy of the cleaning procedures should be evaluated by repeating the trace generated only by the detector, i.e., with the column removed and the column fitting at the detector capped off.

Even if the signal has now been reduced to a satisfactory level, it is important to realize that the original problem may have been a dirty inlet that discharged contaminants to the detector via the column (see Section 7.2). If such a situation exists and is not corrected, both the column and the detector will become recontaminated. The condition of the inlet can be checked by installation of a dummy column (a few meters of clean stainless steel or fused silica tubing will usually suffice). With the detector energized, the high-temperature signal trace should be

---

*For systems with large-diameter open tubular columns operating at flow rates >10 ml/min with nitrogen or helium and employing FID, it may be necessary to add 10–30 ml/min of clean nitrogen at the makeup adapter to supply sufficient gas volume to support the flame with the column removed.

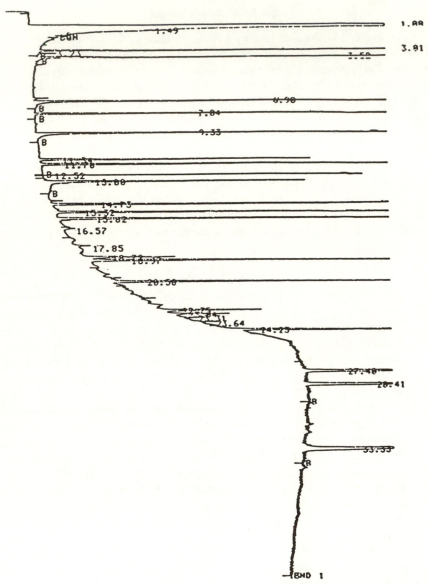

Fig. 10.3. Excessive bleed; although this type or problem can be caused by residues and deposits in the inlet, the detector, or associated lines/filters, in this case it was due to the regeneration of free silanols by oxygen contamination. See text for details.

reevaluated; excessive baseline rise indicates contamination originating from the inlet side. The source may be a dirty injector or deposits in the carrier gas line, flow controllers, regulators, or expended gas scrubbers.

Flashback, which occurs when the abrupt increase in sample volume cannot be accommodated by the volume of the vaporization chamber and the rate at which the carrier gas sweeps it to the column, is usually evidenced by a tailing solvent peak and poor quantitation (Chapter 3; Section 7.7); flashback can also contaminate with residues of everything in that injection any area into which the rapid expansion forces the solvent–solute mixture; this includes the septum and septum seat, the liner housing, and even the carrier gas line and filters therein. Such deposits can lead to degradation of materials in later injections, malformed and tailing peaks, ghost peaks, and bleed. Flashback is more severe (1) with the injection of larger sample volumes, (2) where the volume of carrier gas flowing through the inlet is small relative to the volume of the vaporization chamber, and (3) where the injector temperature is very much higher than the boiling point of the solvent. Again, thorough cleaning is required; inlet liners should be removed and all surfaces and cavities within the injector block should be cleaned, including those that would appear to be out of the main flow stream.

Once a satisfactory trace is generated on the dummy column, attention can be directed to the original column. Some columns with cross-linked surface-bonded stationary phases can be cleaned by rinsing with several column volumes of solvent (see below). In other cases, the column can be connected to the inlet with the detector end disconnected; detector gases should be off and the detector connection plugged. The column should be reconditioned at a relatively high gas flow (unless provisions are made to discharge the carrier outside the oven, hydrogen carrier is not recommended), at a temperature at or approaching that where bleed was previously experienced. Less column damage will usually occur if conditioning is done for longer times at lower temperatures rather than vice versa. Following conditioning, the column should be reconnected to the detector, the detector energized, and the entire system (inlet–column–detector) evaluated by signal trace.

If neither the trace from the detector nor that from the inlet–dummy column–detector exhibited a high-level signal, then the observed bleed must have originated with the column and could have come from several causes. Sources of column bleed include

1. materials dispersed through the stationary phase polymer that become volatile at higher temperatures (usually oligomers that failed to polymerize) and degradation products of the polymer, produced by
2. depolymerization triggered by residual catalyst;
3. reaction with active sites (e.g., $-OH$) on the solid support;
4. reaction with $-OH$ groups within the stationary phase;

5. reaction of the stationary phase with injected materials or injection residues;

6. reaction of the stationary phase with impurities and contaminants (e.g., O_2) in the carrier gas; and

7. the eventual emergence of low-volatility materials deposited on the column from other sources, including higher-boiling solutes that were present in previously injected samples or inadequately cleaned carrier gas.

Any oligomeric materials that are present, either as residues from incomplete polymerization or as products of a depolymerization triggered by residual catalyst, will of course contribute to column bleed, but bleed contributed by factors 1 and 2 is not usually a problem with the higher-quality stationary phases, and the contribution of factor 3 is minimized by careful deactivation procedures; control of these three factors is a responsibility of the column manufacturer. The manufacturer should also have taken steps to minimize bleed due to factor 4 by endcapping the terminal $-OH$ groups of the polysiloxane (Chapter 4). Factor 6 can lead to reappearance of this problem; exposing the column to even traces of oxygen at temperatures above 250°C cleaves $Si-CH_3$ bonds and regenerates $Si-OH$ entities [6]; these engender degradative reactions that are discussed in greater detail below [7–10].

An example of factor 5 occurs with the injection of ethereal extracts of alkaline solutions, which can lead to degradation of phase in the front portion of silicone-coated columns and to severe bleed problems [11].

Residues from the injection of ''dirty'' or ''real-world'' samples accumulate at points of sample vaporization; with properly executed split and splitless injection, these residues accumulate in the inlet liner, which should periodically be inspected, cleaned, and replaced. With on-column injections, injection residues accumulate on the column per se and are commonly evidenced as peak broadening (loss of separation efficiency); other peak-related anomalies such as tailing or splitting have also been attributed to injection residues [11]. Where the residues are of a carbohydrate or proteinaceous character, polar solutes are usually more affected; residues from waxes, plastics, and lubricating oils tend to exercise their greatest effect on apolar solutes [11]. Residues from previous injections can also interact with solutes in subsequent injections vaporized from that soiled surface; degradation can lead to fragmentation, and silyl derivatives can decompose and contribute additional nonvolatile residues [12].

10.4   Temperature and Oxygen Effects

Polysiloxane Stationary Phases

Polysiloxanes exposed to high temperatures and/or oxygen undergo degradation; dimethyl polysiloxane subjected to programmed heating in vacuum produced volatile degradation products, primarily the cyclic trimeric oligomer, which was detectable at 343°C and reached a maximum at 443°C [13]. The reaction, which was accompanied by cross-linking, reportedly occurred in a stepwise manner from the hydroxyl-terminated ends of the polymer chain, was accelerated by oxygen, and was strongly accelerated by KOH. The polymer was much more stable when the terminal hydroxyls were replaced by trimethylsilyl groups (end-blocking). Later work demonstrated that the substitution of phenyl for some of the methyl resulted in a more thermally stable polysiloxane [14]. Larson *et al.* [15] used columns coated with OV-1 (dimethyl polysiloxane) and SE-54 (5% phenyl, 1% vinyl methyl polysiloxane) with hydrogen, helium plus 110 ppm O_2, and helium plus 525 ppm O_2 as carrier gases at 325 and 400°C in their studies of column stress parameters. Their results also indicated that phenyl substitution lent both thermal and oxidative stability to the polysiloxane. Evans [16] measured the retention indices of solutes on packed columns containing several different stationary phases that had been subjected to air flow at 225°C for up to 60 hr. Among the polysiloxanes, the largest changes occurred with the dimethyl siloxane OV-101 and the trifluoropropylmethyl siloxane OV-210; the most resistant to change was the 50% phenyl OV-17, while the cyanopropyl-phenylmethyl silicone OV-225 was intermediate.

Polyethylene Glycol Stationary Phases

The polyethylene glycol-based phases are in general less stable than the poly-siloxanes. It has been suggested that three factors are involved in their decomposition: oxygen in the carrier gas, the catalytic action of the solid support, and catalysis by acidic products of the degradation [17]; among the decomposition products of Carbowax 20M are acetaldehyde and acetic acid [18]. The thermal decomposition of the Carbowax stationary phases is encouraged by even trace levels of oxygen. Conder *et al.* [17] reported that inclusion of an oxygen scrubber that reduced the O_2 content of their commercial carrier gas by three orders of magnitude decreased the rate of Carbowax 20M decomposition by a factor of 5 and raised the temperature at which decomposition was first evidenced from 160 to 200°C.

10.5   Column Rejuvenation

Sources of oxygen contamination were considered in Chapter 7, and the material explored above indicates that polysiloxane-coated columns can be severely damaged by exposure to oxygen at high temperatures (Fig. 10.3). Oxidative cleavage of Si–C bonds regenerates hydroxyl groups on the polysiloxane chain, which engage in "backbiting" reactions to produce cyclic trimers (primarily) of Si–O from the end of the chain. It is this material that generates the "bleed" signal. With each such reaction, a new −OH is left on the chain to perpetuate the reactions. Capping of the exposed hydroxyl groups, which would block those degradative reactions, has been the subject of several studies (e.g., [7–10]); particularly with cross-linked, surface-bonded, high-methyl polysiloxane columns (e.g., DB-1, DB-5), one of the cleaning and resilylation procedures of Grob and Grob [7] has been found especially effective.

The column must first be thoroughly cleaned; otherwise the silylation reactions will confer volatility on injection residues and create a new bleed problem whose correction will be difficult. The column should be rinsed with several column volumes of an appropriate solvent, and it may be desirable to use both polar and nonpolar solvents. Depending on the nature of the sample residues expected, the first rinse might be water, followed by methanol and acetone; methylene chloride is a good final rinse and in some cases may be the only solvent required. The column should then be filled with methylene chloride and allowed to stand flooded overnight to allow materials within the stationary phase time to migrate into the solvent. The column is flushed with fresh methylene chloride, drained, and dried at room temperature with a stream of clean dry nitrogen. The heavily solvated stationary phase may now appear as a thick puffy film; at temperatures above the boiling point of the solvent, violent solvent evaporation can tear the film to pieces. Instead, the column should be subjected to a nitrogen flow stream and held at a temperature just below the boiling point of the solvent for several hours; the temperature should then be increased at a very low rate to a point perhaps 20°C above the boiling point of the solvent, and held for 2 or 3 hr. After allowing the column to cool, vacuum is used to draw a 10% solution of diphenyltetramethyldisilazane (DPTMDS) in pentane into the column until about one-tenth of the column is filled. A very slow flow of clean dry nitrogen is passed through the column to encourage evaporation of the ether; the flow should not be so fast that plugs of liquid are blown from the end of the column. When the ether appears to have evaporated, the nitrogen flow is increased slightly (2–3 ml/min for 0.25–0.32-mm columns; 10–20 ml/min for 0.53-mm columns) and the column is held at about 75°C to ensure total evaporation of the solvent (approximately 1 hr), leaving the DPTMDS in the column. Both ends of the column are then flame-sealed in such a manner as to ensure that it is nitrogen-filled, and the column is heated to 300°C for 30–60 min, cooled, flushed with methylene chloride, and dried.

Polysiloxane phases containing higher levels of phenyl or phenylcyanopropyl give variable results on resilylation; improvements may occur in the bleed level of 1701-type phases, but aggressiveness toward active solutes (e.g., alcohol tailing) has been noted [19]. Berezkin and Korolev [20] reported that the contribution of adsorption to the retention behavior of polar solutes increases as the degree of cross-linking increases, and resilylation could result in more extensive cross-linking of the stationary phase.

10.6 Peak Distortion

Injection technique can affect peak conformation and in extreme cases result in misshapen, malformed, or split peaks (Chapter 3); the relative polarities of solvent, solutes, and stationary phase have also been shown to play roles [21,22]. Other general causes of peak distortion and asymmetry have been addressed by Conder [23], who described the two extremes in peak skewing as (1) a simple skewing of the whole peak without a long tail and (2) tailing (or "fronting") where the (very) upper part of the peak is nearly symmetrical, but is followed (or preceded) by a distinct, drawn-out tail of exponential form.

Fronting, where the leading edge exhibits a slow rise and the drop from the peak maximum is much more precipitous, is generally due to "vapor phase overload," in which the solute vapor pressure (dictated by the column temperature) is too low to accommodate in the vapor phase the amount of that solute required to maintain the equilibrium K_D [24]; a typical example is shown in Fig. 10.4. The defect can be corrected by decreasing the sample size (which would

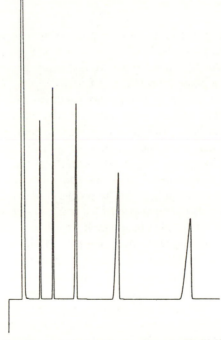

Fig. 10.4. "Fronting" peaks, typical of a "vapor phase overload" [24]. In this example, the phenomenon is apparent with peak number 3 (excluding solvent) and becomes progressively worse with decreasing vapor pressure of the solutes.

Fig. 10.5. The "Christmas tree" effect; peak splitting artificially generated in *n*-pentadecane by inducing short-term temperature fluctuations in the detector limb of a low thermal mass column [27].

decrease the amount of that solute in both phases), increasing the column temperature (which would increase the solute vapor pressure), or increasing the solute K_D {by substituting a more retentive stationary phase ($K_D = c_S/c_M$) or increasing the column phase ratio [$K_D = \beta k = (r/2d_f)k$]}.

Skewing that takes the form of tailing may result from flashback (see above), from gas flow problems, or from active sites that may be present in the inlet, the column, or the detector itself. Tailing of hydrocarbon peaks generally testifies to flashback or to inadequacies in the gas flow stream; the split ratio may be too low, or the flow of makeup gas may be low or misdirected. Activity in the flow stream is indicated if hydrocarbons yield well-formed peaks while active solutes exhibit tailing; column substitution can help determine whether the problem is extracolumn. The inlet should be inspected for residues and cleaned if necessary, and positioning of the outlet end of the column should be verified; deactivation of the inlet liner may be necessary. If the problem seems to be associated with the column, it may help to backflush the column with solvent to remove residues. The removal of one to several meters of column from the inlet end may also correct the problem; because resolution varies with the square root of column length, shortening the column rarely has a pronounced effect on the separation efficiency. Column resilylation may correct activity problems in some cases, but in other cases it may accentuate tailing (see above).

Peak splitting and "Christmas tree" effects are discussed in Chapter 7; these interrelated problems are usually attributed to exposure of a portion of the low thermal mass columns to nonconstant thermal gradients or radiant heat sources

[25–28] and are related to acceleration of restricted portions of the chromato-graphing band. Reed and Hunt [29] studied the phenomena by subjecting the detector limb of the column to a pulsating heat source (Fig. 10.5).

10.7 Other Sorptive Residues

Any of the phenomena discussed above (split and malformed peaks and mem-ory effects that result in ghost peaks) can also be caused by chromatographic conditions that permit the sample to come in contact with materials that are capable of interacting with sample components; these interactions can range from adsorption to solution. Chapter 7 discussed risks associated with graphite parti-cles that are scraped from the ferrule by the column end to lodge in the flow stream; the series of ECD chromatograms shown in Fig. 10.6 illustrates a related problem [30]. Analysis I shows the programmed temperature separation of a series of pesticides; analysis II was run 1 hr later and exhibits peak distortions. Analysis III is the same mixture, injected immediately after analysis II. In IV, V, and VI, injections of the solvent only (methanol) establish that these anomalies are attributable to memory effects (ghosting). In analysis VII the column was programmed without injection, and VIII was run 15 min after VII; both indicate a source of sample in the flow stream. A postcolumn source would yield a high background signal, but the components would not be separated. Because the "peaks" exhibit retentions comparable to those exhibited by normal injections, the source must precede the column or be situated in the very first section of the column. Careful examination of the injector block with the liner removed re-vealed that a minute crumb of silicone rubber, presumably chewed from the septum by the syringe needle, had lodged within the carrier gas line at the point of entry to the injector. The bottom trace shows the baseline generated under programmed conditions after removal of that small particle. Other sorptive mate-rials in the flow stream can give rise to entirely different problems (see below).

10.8 Column Coupling and Junction Problems

Some type of junction is required in the attachment of a retention gap to the column, in coupling columns, or where the column effluent is split to two detectors. Commercial unions and tees can be satisfactory under conditions of slow program rates or isothermal operation. With faster program rates, the appre-ciable thermal mass of the fitting may result in its temperature lagging behind that of the column and the oven, and higher-boiling solutes can be cold-trapped in the fitting. A homologous series of hydrocarbons can exhibit good conforma-tion of earlier peaks, but a sudden break in their elution sequence signals trouble; as the temperature of the fitting slowly increases, trapped solutes are eluted as split and malformed peaks; "Christmas tree" effects may also be apparent. One

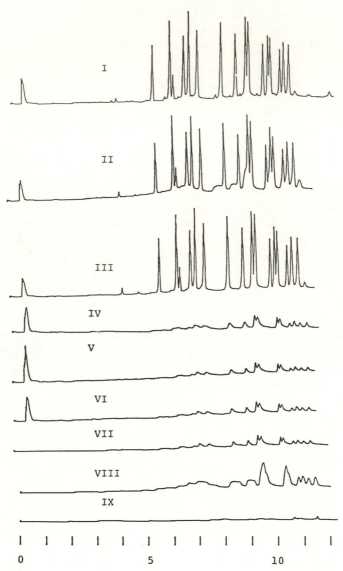

Fig. 10.6. Effects of sorptive material in the inlet; 30 m × 0.53 mm column, programmed runs; ECD. (I) Normal separation, (II) second analysis immediately after completion of I, (III) analysis after a 1-hr pause, (IV–VI) methanol injections, (VII, VIII) column programmed without injection, (IX) programmed baseline following removal of a small particle of silicone rubber from the inlet. See text for discussion.

solution is to wire the metal fitting tightly against the base of the heated detector; another is to accelerate transfer of heat into the high thermal mass fitting by enclosure in a separately heated "oven" [31].

Junctions imparting a lower thermal mass have also been explored; ends of the two pieces of tubing can be slipped into a close-fitting glass sleeve and sealed with polyimide resin. After careful air-drying, the resin is cured by heating the column under gas flow; the flow must be sufficient to protect the column during heating but not so great that the polyimide sealing the junction "blows" during the cure. An additional danger is the possibility of the absorptive polyimide entering the flow stream (Fig. 10.7).

More permanent junctions may justify fusion couplings [32], in which the ends of the two tubes are stripped of polyimide and fused within a borosilicate

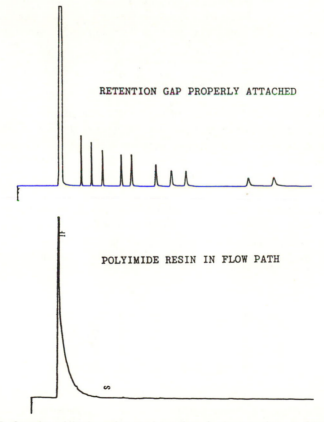

Fig. 10.7. On-column injections of a test mixture through a 1-m retention gap of deactivated fused silica tubing, attached to the column by means of a polyimide-sealed glass sleeve. Top, normal results; bottom, polyimide in flow stream.

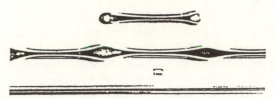

Fig. 10.8. Stages in the production of glass press-fit connectors [33]. The ends of the fused silica tubing must be cut squarely and seated firmly into the tapered junction. Both radial symmetry and a gradual taper are essential to the success of this coupling device.

glass sleeve; polyimide resin can then be applied over the junction to restore stress resistance. Press-fit connectors that have recently been described seem promising (Fig. 10.8) [33]; the same concept was employed in sealing large-diameter open tubular columns into an inlet designed to minimize flashback problems (Chapter 4). Thin-walled glass sleeves with ground tapers that seal with columns of specific outer diameter are currently under investigation and show some promise.

10.9   Flame Jet Problems

The small diameter and flexibility of the fused silica column can be used to good advantage at the detector end, because the column end can be sited so that solutes are eluted directly at the point of detection, eliminating the possibilities of adsorptive sites in transfer tubing, flame jets, etc. Some of the problems that can occur are illustrated in Fig. 10.9, which shows a fused silica column terminating within a flame jet. If the diameter of the column exceeds (I) or is similar to (II) that of the jet orifice, the flow of combustion hydrogen (and makeup gas) may be restricted to such a degree that the flame cannot be ignited or maintained. Thrusting a smaller column completely through the jet (III) exposes the column to the flame; this condition is usually characterized by wild, fluctuating signals

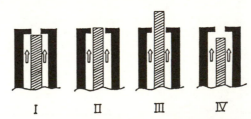

I   II   III   IV

Fig. 10.9. Positioning of the outlet end of the column within the flame jet. Configurations I and II restrict the flow of gases through the orifice and make difficult ignition and maintenance of the flame; configuration III generates wild signals, due to the presence of the column in the flame. The ideal configuration is as shown in IV.

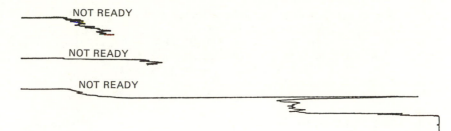

Fig. 10.10. Signal generated by an FID with the column end positioned as in III, Fig. 10.9.

(Fig. 10.10). Ideally, the column should terminate 2–4 mm below the jet orifice (IV), in such a position that solute contact with the flame jet is precluded by the flow path of the surrounding gases and the end of the column does not restrict free passage of gas from the jet orifice.

References

1. W. Jennings, "Gas Chromatography with Glass Capillary Columns," 2nd ed. Academic Press, New York, 1980.
2. "Troubleshooting Guide," Supelco Guide 792. 1983.
3. K. Grob, G. Grob, and K. Grob, Jr., *J. Chromatogr.* **156,** 1 (1978).
4. W. Jennings, "Comparisons of Fused Silica and Other Glass Columns in Gas Chromatography." Huethig, Heidelberg, 1981.
5. I. Temmerman and P. Sandra, *J. High Res. Chromatogr.* **7,** 332 (1984).
6. A. E. Coleman, *J. Chromatogr. Sci.* **13,** 198 (1973).
7. K. Grob and G. Grob, *J. High Res. Chromatogr.* **5,** 349 (1982).
8. g. Schomburg, R. Dielmann, H. Borwitzky, and H. Husmann, *J. Chromatogr.* **167,** 337 (1978).
9. V. Paramasigamani and W. A. Aue, *J. Chromatogr.* **168,** 202 (1979).
10. S. Hoffmann, L. G. Blomberg, J. Buijten, K. Markides, and T. Wannman, *J. Chromatogr.* **302,** 95 (1984).
11. K. Grob, Jr., *J. Chromatogr.* **287,** 1 (1984).
12. M. Donike, *Chromatographia* **6,** 190 (1973).
13. N. Grassie and I. G. MacFarlane, *Eur. Polym. J.* **14,** 875 (1978).
14. N. Grassie, I. G. MacFarlane, and K. F. Francie, *Eur. Polym. J.* **15,** 415 (1979).
15. P. Larson, T. Stark, and R. Dandeneau, *Proc. Int. Symp. Capillary Chromatogr., 4th, 1981,* p. 727 (1981).
16. M. B. Evans, *Chromatographia* **15,** 355 (1982).
17. J. R. Conder, N. A. Fruitwala, and M. K. Shingari, *J. Chromatogr.* **269,** 171 (1983).
18. J. Debraurwere and M. Verzele, *J. Chromatogr. Sci.* **14,** 296 (1976).
19. W. Jennings, *J. Chromatogr. Sci.* **21,** 337 (1983).
20. V. G. Berezkin and A. A. Korolev, *Chromatographia* **20,** 482 (1985).
21. R. G. Jenkins, *Proc. Int. Symp. Capillary Chromatography, 4th,* p. 803, *1981* (1981).
22. L. Ghaoui, F. S. Wang, H. Shanfield, and A. Zlatkis, *J. High Res. Chromatogr.* **6,** 497 (1983).
23. J. R. Conder, *J. High Res. Chromatogr.* **5,** 341, 397 (1982).
24. K. Yabumoto, D. F. Ingraham, and W. Jennings, *J. High Res. Chromatogr.* **3,** 248 (1980).

25. S. A. Mooney, *J. High Res. Chromatogr.* **5,** 507 (1982).
26. G. Schomburg, *J. Chromatogr. Sci.* **21,** 97 (1983).
27. F. Munari and S. Trestianu, *J. Chromatogr.* **279,** 457 (1983).
28. F. J. Schwende and J. D. Gleason, *J. High Res. Chromatogr.* **8,** 29 (1985).
29. G. D. Reed and R. J. Hunt, *J. High Res. Chromatogr.* **9,** 341 (1986).
30. D. Kukla, personal communication (1986).
31. M. F. Mehran, W. J. Cooper, M. Mehran, and F. Diaz, *J. High Res. Chromatogr.* **7,** 639 (1984).
32. K. Grob, Jr., *J. Chromatogr.* **330,** 217 (1985).
33. E. R. Rohwer, V. Pretorius, and P. J. Apps, *J. High Res. Chromatogr.* **9,** 295 (1986).

ABBREVIATIONS, TERMS, AND NOMENCLATURE

Terms and symbols for gas chromatography have frequently been coined as required by the many individual contributors; as a result, different symbols and definitions are often applied to the same parameter. The resulting confusion will continue until a unified system gains general acceptance. Systems for gas chromatographic nomenclature have been proposed by two established organizations: the International Union of Pure and Applied Chemistry (IUPAC) [1] and the American Society for Testing and Materials (ASTM) [2].

Although there are many similarities between the two systems, there are also some notable differences. As one example of the latter, the ratio of the concentrations of a solute in stationary and mobile phases is termed the "distribution constant" (and denoted by the symbol K_D) by IUPAC, while ASTM uses the term "partition coefficient" (and employs the symbol K) for this parameter. Another difference occurs in the use of the term "partition ratio" by IUPAC for the ratio of the amounts of a solute in stationary and mobile phases, respectively, versus the ASTM "capacity factor."

The IUPAC system predates that of ASTM and has been used in earlier publications by this author. The terms listed below are restricted to those used in this book and include abbreviations widely used in gas chromatography, plus nomenclature terms. Many of the latter are common to both systems; where the two systems differ, preference has been given to that based on terms or usage which extensive teaching experience indicates most students find easier to grasp. Accordingly, the IUPAC "partition ratio" is used in preference to the ASTM "capacity factor"; it should also be noted that both systems recommend the symbol k rather than the archaic k'. If we accept this dual recommendation for the symbol k, the easier comprehension of the oral expression "$K_D = \beta k$" as compared to "$K = \beta k$" also mandates the IUPAC symbol K_D for our purposes here.

A Packing factor term of the van Deemter (packed column) equation.

α, **relative retention**   Ratio of the adjusted retention times of any two solutes, measured under identical conditions. Because α is never less than 1.0, it is the function of the more retained solute relative to the same function of the less retained solute:

$$\alpha = t'_{R(B)}/t'_{R(A)} = K_{D(B)}/K_{D(A)} = k_{(B)}/k_{(A)}$$

This parameter is termed the "separation factor" by IUPAC.

B Longitudinal diffusion term of the van Deemter (packed column) and Golay (open tubular column) equations.

β, **Column phase ratio**   Volume of column occupied by mobile (gas) phase relative to the volume occupied by stationary (liquid) phase. In the open tubular column,

$$\beta = (r - 2d_f)/2d_f \simeq 0.5r/d_f$$

The use of d instead of r and of different units for r and d_f are common student errors.

c_M, c_S Solute concentrations in mobile and stationary phases, respectively.

C Resistance to mass transport (mass transfer) term in the van Deemter equation; C_M and C_S denote mass transport from mobile to stationary and from stationary to mobile phases, respectively.

d Column diameter. Both millimeters and micrometers are commonly used; the latter, while consistent with the units used for d_f, implies an accuracy to three significant figures, which is rarely correct.

d_f Thickness of the stationary phase film, usually in micrometers.

D Diffusivity; D_M and D_S represent diffusivities in mobile and stationary phases, respectively.

ECD   Electron capture detector.

f_1, f_2 Gas compressibility (pressure drop) correction factors.

F Volumetric flow of the mobile phase, usually cubic centimeters per minute. IUPAC uses the same symbol for "nominal linear flow" and F_c for volumetric flow; ASTM uses F_a for "gas flow rate from column."

FID   Flame ionization detector.

FPD   Flame photometric detector.

GC/MS   Gas chromatography/mass spectrometry.

h Length of column equivalent to one theoretical plate. Less precisely termed "height equivalent to a theoretical plate" (which accounts for the ASTM symbol "HETP"), this value is obtained by dividing the column length by the theoretical plate number and is usually expressed in millimeters:

$$h = L/n$$

When measured at $\bar{u}_{opt}$, the result is termed h_{min}.

H   Length of column equivalent to one effective theoretical plate:

$$H = L/N$$

i.d.   Inner diameter of column.

k, partition ratio   Ratio of the amounts of a solute in stationary (liquid) phase and mobile (gas) phase, which is equivalent to the ratio of the times the solute spends in the two phases. Because all solutes spend time t_M in mobile phase,

$$k = (t_R - t_M)/t_M = t'_R/t_M$$

ASTM prefers the term "capacity ratio"; neither nomenclature uses the symbol k', which must be considered archaic.

K_D, distribution constant   Ratio of the concentrations of a solute in stationary and mobile phases:

$$K_D = c_S/c_M$$

ASTM uses the symbol K and the term "partition coefficient" for this parameter.

L   Length of the column.

n   Theoretical plate number:

$$n = (t_R/\sigma)^2$$

where σ is the standard deviation of the peak.

n_{req}   Number of theoretical plates required to separate two solutes of a given α and given partition ratio to a given degree of resolution.

N   Effective theoretical plate number:

$$N = (t'_R/\sigma)^2$$

ASTM refers to this parameter as the "effective plate number"; the IUPAC nomenclature recognizes that those plates are still theoretical.

NPD   Nitrogen/phosphorus detector.

o.d.   Outer diameter of the column.

r   Radius of the column.

R_s, peak resolution   Measure of separation as evidenced by both the distance between the peak maxima and the peak widths. Both IUPAC and

ASTM definitions are based on w_b measurements, which must be determined by extrapolation. If the peaks are assumed to be Gaussian,

$$R_s = 1.18[(t_{R(B)} - t_{R(A)})/(w_{h(A)} + w_{h(B)})]$$

σ Standard deviation of a Gaussian peak.

t_M, **gas holdup time**   Time (or chart distance) required for elution of a nonretained substance (e.g., mobile phase). "Holdup time" and "holdup volume" are preferred to the terms "dead time" and "dead volume."

t_R, **retention time**   Time (or distance) from the point of injection to the point of the peak maximum.

t'_R, **adjusted retention time**   Equivalent to the residence time in stationary phase, this is determined by subtracting the gas holdup volume from the solute retention time:

$$t'_R = t_R - t_M$$

This should not be confused with the term "corrected retention time," which is defined differently.

$\bar{u}$ Average linear velocity of the mobile phase:

$$\bar{u} = L(\text{cm})/t_M(\text{sec})$$

v Linear velocity, usually of the solute band.

V Volume; V_M and V_S represent the volumes of mobile and stationary phases, respectively.

w_b, **peak width at base**   This is determined by measuring the length of baseline defined by intercepts extrapolated from the points of inflection of the peak. Equivalent to four standard deviations (σ) in a Gaussian peak.

w_h, **peak width at half height**   Measured across the peak halfway between the baseline and the peak maximum, this can be determined directly without extrapolation. In a Gaussian peak, equal to 2.35 standard deviations (σ).

INDEX